'08

Food for Thought

ISSUES
(previously Issues for the Nineties)

Volume 36

Editor

Craig Donnellan

Independence

First published by Independence
PO Box 295
Cambridge CB1 3XP
England

British Library Cataloguing in Publication Data
Food for Thought – (Issues Series)
I. Donnellan, Craig II. Series
613.2

ISBN 1 86168 058 9

Printed in Great Britain
City Print Ltd
Milton Keynes

Typeset by
Claire Boyd

Cover
The illustration on the front cover is by
The Attic Publishing Co.

CONTENTS

Introduction

Food for Thought is the thirty-sixth volume in the series: **Issues**. The aim of this series is to offer up-to-date information about important issues in our world.

Food for Thought looks at the average diet, eating habits and food safety issues.

The information comes from a wide variety of sources and includes:
Government reports and statistics
Newspaper reports and features
Magazine articles and surveys
Literature from lobby groups
and charitable organisations.

It is hoped that, as you read about the many aspects of the issues explored in this book, you will critically evaluate the information presented. It is important that you decide whether you are being presented with facts or opinions. Does the writer give a biased or an unbiased report? If an opinion is being expressed, do you agree with the writer?

Food for Thought offers a useful starting-point for those who need convenient access to information about the many issues involved. However, it is only a starting-point. At the back of the book is a list of organisations which you may want to contact for further information.

*****

Children's views on food and nutrition

Information from a survey by the European Food Information Council (EUFIC)

Current eating and drinking patterns

Children were asked to give details of one of their meals during the last 24 hours preceding the interview.

Typical menus for breakfast, lunch, and dinner in these four countries show that interesting similarities exist between diets in France and Italy, as between those in Germany and the UK.

Breakfasts in France and Italy consist mostly of bakery products; likewise for Germany. In the UK, breakfast cereals top the ranking of the most frequently consumed breakfast foods, and they are in second place in Germany. When it comes to breakfast drinks, French children favour hot chocolate, Italians and Germans prefer milk, and British children drink tea the most frequently.

Almost one in five (19%) children in Italy does not eat anything for breakfast, a percentage which is much higher than in the other countries; in the UK it is 9%, in France, 5%, and just 1% in Germany.

At lunch, vegetables and salad are quite popular across all four countries, with over half of the children surveyed eating them. French and Italian children consume fruit much more frequently than children in Germany and the UK. French children eat yoghurt much more frequently than children elsewhere. Amongst carbohydrate-rich foods, pasta is popular in France and Italy, while potatoes are favoured in Germany and the UK. British children eat french fries much more often than do their counterparts in the other countries. Sandwiches are particularly popular in the UK, where almost one-quarter of children eat them for lunch, compared with less than 6% in the other three countries. When it comes to protein-rich foods, French and Italians eat meat, chicken, fish, and cheese more often than children in the other two countries do.

Water is the most popular lunchtime drink in France and Italy, with a variety of colas and lemonades being favoured in Germany and the UK.

The dinner menus of school-children include a diverse range of foods. Amongst carbohydrate-rich foods, pasta is favoured in France and Italy, while German children prefer bread, and British children eat potatoes, especially french fries, more often. Of protein-rich foods, children in France, Germany, and the UK eat meat more frequently than do children in Italy; French and Italian children consume fish more frequently than the others; and German children eat cheese more often.

Children in France and Italy favour water as a drink with dinner, whilst children in Germany and the UK favoured a greater variety of drinks.

Children claim to have a significant influence on the choice of menu at breakfast. Across the four countries studied – France, Germany, Italy and the UK – 70% claim to make their own decisions on what to eat during the week, even though 48% eat this meal with their family. For the evening meal – both during the week and at weekends – only 28% claim to decide for themselves what to eat. School routines influence the decision-making of children at lunch time.

Approximately half of children surveyed in France, Germany and Italy eat their weekday breakfast with one or more parents. In the UK, 36% share this meal with parents. At weekends, around two-thirds of children in France and Italy eat breakfast with parents, compared

with less than half of UK children and over four-fifths (82%) of German children.

During the school term, two-fifths (41%) of children in France, 56% in Germany, 77% in Italy and only 4% in the UK eat lunch with at least one parent. At the weekend, over 90% of children in France, Germany and Italy share lunch with a parent, compared with 58% in the UK.

Across all four countries studied, the vast majority of children eat the main evening meal with one or more parents. In France, 96% of children have their evening meal with the family both during the week and at weekends; in Germany, 88% eat as a family during the week, a figure which rises to 92% at the weekend; in Italy, 96% eat with parents on weekdays and 95% at the weekend; and in the UK, 87% of children eat with parents during the week, which rises to 91% at weekends.

Children's views on nutrition and health

Overall, children are pretty well-versed in understanding nutrition and recognising the importance of a balanced diet for good health. The majority of children agree that 'Milk is good for strong bones' (84% agree); 'Foods like sweets and ice cream are OK to eat, but not all the time' (82%); 'To stay healthy, you should eat less fat' (82%); 'The food you eat affects your health as you grow up' (81%); 'Exercise is just as important as the foods you eat for staying healthy' (79%); 'It is important to eat foods like whole grain bread and cereals' (79%); and 'It is best to eat small amounts of different foods, rather than a lot of the same food' (74%). Only 11% agree that 'Chocolate is OK to eat every day', and 8% agree that 'Fast foods are OK to eat every day.'

Across the four countries, on average, children rate the importance of nutrients to health as follows: vitamins (94% of children agree that the body needs them in order to stay healthy), protein (82%), calcium (78%), minerals (70%), fibre (61%), calories (43%), sugar (42%), starch (38%), salt (34%), fat (33%) and pills (7%).

When asked what foods are 'Good for you', the most frequent answer is fruits (85%), followed by vegetables (82%), water (73%) and milk (72%). The four-country 'Not so good for you' ranking is led by beer (78%), wine (74%), sweets (69%) and cider (66%). Minor differences in these rankings were recorded in each country.

When asked to compare the nutritional value of fresh foods and processed foods, 61% disagree with the statement, 'Ready-made meals are just as good for you as home-made meals' and 60% disagree that 'Tinned fruit and vegetables are as good for you as fresh fruit and vegetables'. Almost one in six is unable to reply. Some 69% of children believe that 'Fresh foods are safer than tinned or frozen foods.'

Learning about food and nutrition

Most children understand that the word 'nutrition' is related to the health and dietary benefits of foods. However, close to one-third of children in France (32%) and the UK (34%) were unable to express any opinion on the word's meaning. This figure was much lower in Germany (6%) and Italy (3%).

Across all four countries, an average of 92% of children believe that it is important to learn about nutrition. Only 7% of children claim to 'know a lot' about nutrition. A further 56% 'know some things', 30% 'don't know much' and 7% 'don't know anything' about nutrition.

Overall, the family is perceived by 67% of children as being their most important source of information on nutrition. The child's school (41%) and school teachers (34%) are also perceived as playing a central role. Television programmes are a source of information for 17% of children. Children less often learn about nutrition from advertising, magazines, and friends.

Half of all children agree that cooking classes at school would be the best way to learn about nutrition. Some 43% would prefer information packs for use at school, whilst 31% would like to learn more from television programmes.

Food safety and hygiene

Overall, children appear to understand the basic rules of food hygiene. The majority of children agree with the statements: 'You should always wash fruit and vegetables before eating them' (96% agree); 'It is important to keep the kitchen clean' (95%); 'Food can go "off" from being kept at the wrong temperature' (86%); and 'You should always cover food before putting it back in the fridge' (83%).

Some 77% of children are aware that 'use-by' dates on food mean that 'You need to eat the food before that date', whilst 8% say that they do not know what 'use-by' dates mean.

When asked whether 'Foods containing E numbers are bad for you', 58% of children claim they do not know. 42% are in agreement with the statement 'The food we eat today is safer than ever before'; 31% do not know whether this is the case. 53% disagree with the statement, 'It is safer to drink milk straight from the cow than the milk one buys in the store'; 25% claim they do not know whether this is true.

Concerning food poisoning, 61% of children believe that it is caused by 'Harmful bacteria in food', whilst 28% attribute it to 'The pesticides that are used on crops'. Some 49% are aware that Salmonella food poisoning is caused by harmful bacterial in food, while 41% do not know what Salmonella is.

• The above is an extract from *Children's Views on Food and Nutrition: A Pan-European Survey*, produced by the European Food Information Council (EUFIC). See page 41 for address details.

© EUFIC
October, 1995

Growing up green

Why do so many children grow up disliking fruit and vegetables? Heather Welford on the alarming effects of a bad diet on development

Britain has a problem with food – that's stating the obvious after years when BSE and E-coli have battled for headlines with salmonella and Princess Di's bulimia.

But the more dramatic revelations overshadow a growing problem among children. It seems that many are not getting enough to eat – or not enough of the right sorts of food.

The School Milk Campaign recently reported that up to two million children may be malnourished – their evidence came from a survey of LEAs and health authorities from all over the country. The Community Practitioners and Health Visitors Association reported that 61 per cent of a survey of 500 health visitors had children with iron deficiency on their list, and 4 per cent of the sample had seen cases of rickets (caused by vitamin D deficiency). Up to 80 per cent of them said they saw children who had been diagnosed with 'failure to thrive' – a term used to describe very underweight children.

Last week a team from Strathclyde University found that children aged between three and 16 commonly had an 'arbitrary and despotic dislike of vegetables'. Many children ate the recommended five portions of fruit and vegetables only at Christmas.

The National Diet and Nutrition Survey – part of an ongoing government project – found that pre-school children on average were not going short of nutrients and were growing well. However, within the sample of almost 2,000 children, blood tests showed that some children are missing iron (with 10-16 per cent going short) and zinc (14 per cent missing out). One in 12 children was anaemic.

'Marginal deficiency of these and other minerals may not be spotted unless it's tested for,' says Susan Fairweather-Tait. 'Severe iron deficiency can cause developmental delay, growth retardation and poor language and motor skills. We don't know if these ill effects are completely reversible, either – and we don't yet

> *Within a sample of almost 2,000 children, blood tests showed that some children are missing iron and zinc. One in 12 children was anaemic*

know what effect a slight or temporary deficiency may have.'

Giving supplements of iron or zinc is usually not the answer. 'A few babies or toddlers may need a short course of supplementation once they have been diagnosed as at risk,' she says. 'But the majority should be able to get what they need from their diet. Vitamin C in fruit and vegetables increases the body's ability to use and store iron from a meal and the best way of making sure iron is in the body is to give meat and fish.'

It's not just poor families on a deprived diet whose children are at risk. Failure to thrive appears across all social classes. Dr Rony Waterston, a consultant community paediatrician in Newcastle, confirms the existence of 'muesli-belt malnutrition' in children from middle

Lunch

The West Midlands Young People's Lifestyle Survey is the largest study of its kind and collected information from over 27,000 children between the ages of 11 and 16. The graph below shows that school meals and packed lunches account for the majority of midday meals. Only 17% of the sample considered the midday meal to be their main meal on a school day.

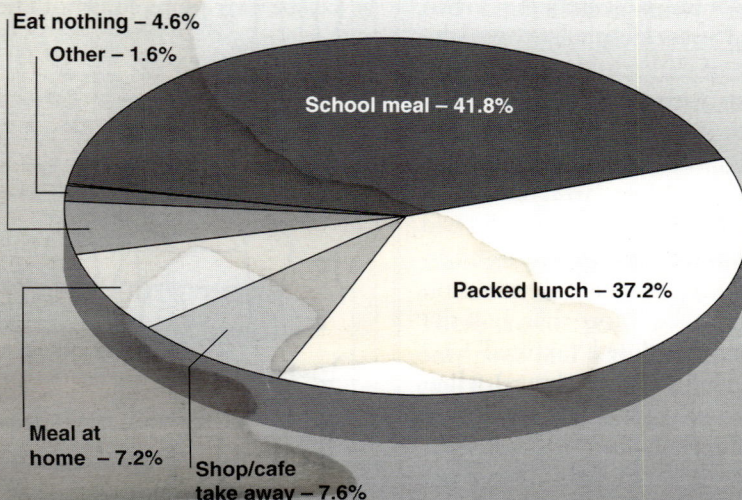

- Eat nothing – 4.6%
- Other – 1.6%
- School meal – 41.8%
- Packed lunch – 37.2%
- Meal at home – 7.2%
- Shop/cafe take away – 7.6%

Source: University of Birmingham

Current eating and drinking patterns

What children eat and drink

Question: *I want to talk to you about food and nutrition, but, first of all, can you tell me what you had to eat and drink for breakfast today/lunch today/ evening meal yesterday? (The child was asked to select one from the previous 24 hours.)*

Key findings:

Breakfast

Bakery products are the most frequently eaten breakfast foods in France, Italy and Germany, while cereals dominate the menu of many UK children (57%), although toast is also popular (26%) there. In none of the four countries surveyed are yoghurt or fruit frequent breakfast choices. Tea is a frequent breakfast drink for British (26%) and German (21%) children; but consumed much less often in the other countries. Italian children drink coffee much more frequently (17%) than do children in the other three countries. Milk, either hot or cold, is the most universal of the breakfast drinks. 48% of French children drink hot chocolate with their breakfast.

In Italy, a significant percentage of children (19%) do not eat anything for breakfast. This compares to 9% in the UK, 5% in France and just 1% in Germany.

Lunch

Vegetables and salad are very popular for lunch in each country – on average, 55% of children eat them for lunch. Potatoes are also eaten by many in Germany (28%) and especially the UK (41%), and less in France (25%) and Italy (11%). In the UK, 30% have french fries at lunch while almost half of Italian children (46%) eat pasta. These two figures are much higher than in other countries.

Of protein-rich foods, meat is the most popular lunch-time choice. 56% of French children eat meat for lunch, which means that they

---

Almost a quarter (24%) of UK children eat sandwiches for lunch, which compares to under 6% for France and Italy and none for Germany

---

consume it more often than children in other countries. Chicken and fish are not very popular lunch-time choices in any country.

Relatively few children in all countries have soup for lunch; the average is only 3%. Almost a quarter (24%) of UK children eat sandwiches, which compares to under 6% for France and Italy and none for Germany. Italy and France have a much higher consumption rate for fruit (30% and 24% respectively) than the UK at 5% and Germany at just 2%. 16% of French children eat yoghurt for lunch, while only 4% of British children do, and 0% of German and Italian children have it for lunch.

The vast majority of children in France (74%) and Italy (90%) drink water; just 21% in Germany and 15% in the UK do so. On average, a number of children choose colas (14%) and lemonades (11%). The choice of cola is especially prominent in the UK (22%).

Dinner

Meat retains its popularity at dinner time in France (44%), Germany (45%) and the UK (44%), but fewer children have it in Italy (24%). 37%

HE LOVES CEREALS & SERIALS FOR BREAKFAST!

KenPyne

of German children have sausages for dinner, and 56% of British children eat potatoes. These two figures are considerably higher than comparable percentages in the other countries. French fries are a popular (34%) dinner choice for children in the UK.

Some country-specific dinner trends are evident. Pasta appears on 27% of dinner tables in Italy, compared to 19% in France, 6% in the UK, and 5% in Germany. Almost half (46%) of German children eat bread at dinner, whereas the four-country average is 19%. 30% of French children have yoghurt at dinner, compared to the four-country average of just 9%.

Vegetables and salad are most popular in the UK (68%) followed by France (43%), Italy (39%) and Germany (27%). More children in the UK (12%) and Italy (10%) have

chicken than in Germany (3%) and France (2%), while fish is more popular in Italy (16%) and France (13%) than in the UK (8%) and Germany (4%). 23% of German children have cheese at dinner, the highest percentage for this product of the four countries (France 17%, Italy 14% and the UK 4%). The popularity of soup at dinner is higher than at lunch, but is still low, with an average of just 7%.

Fruit consumption is again much higher in Italy and France, with 26% of Italian children eating it and 14% of the French. This compares to just 7% of UK children and only 3% of German children.

71% of children in France have water. This compares to 81% in Italy, 14% in the UK and 9% in Germany. Tea is quite popular in Germany (15%) and the UK (14%), as is cola (17% for Germany; 12% for the UK).

In both Germany and the UK, a wide variety of drinks appear at dinner, including lemonade, milk and fruit juice. In the UK, this list is complemented by squashes of assorted flavours.

Methodology footnote:
Each child was asked to give details of one meal eaten during the 24 hours before the interview. He or she was able to select from 'breakfast today/lunch today/evening meal yesterday'; interviewers kept check that overall, a balance of coverage between the different meals was obtained in each country.

• The above is an extract from *Children's Views on Food and Nutrition: A Pan-European Survey*, produced by the European Food Information Council (EUFIC). See page 41 for address details.

© EUFIC
October, 1995

Children's views on nutrition and health

This pan-European study concerning children's views on food and nutrition was commissioned to obtain a basic understanding of the perceptions of European children on food- and drink-related issues. Research was focused on four countries, France, Germany, Italy and the UK. These countries – which have the largest populations of all EU member states – were selected in order to study similarities and differences in dietary habits across Europe.

Which meals do you normally eat with your family?
Dinner is the meal eaten most frequently with the family, both during the week and at weekends. Lunch, particularly at weekends, is also often eaten with one or both parents. Breakfast is eaten with the family most frequently at weekends, although almost half of children across the four countries eat with one or both parents during the week.

%	Breakfast W/D	W/E	Lunch W/D	W/E	Dinner W/D	W/E
(F)	49	64	41	94	96	96
(G)	48	82	56	94	88	92
(I)	58	68	77	96	96	95
(UK)	36	37	4	58	87	91
*	48	65	45	86	92	94

W/D = weekday, W/E = weekend

Statement: Ready made meals are just as good for you as home-made meals
Across the four countries, the majority of children (61%) disagree with the statement. 'Ready-made meals are just as good for you as home-made meals'. Just under one-quarter agree with the statement and about one-sixth don't know.

%	Agree	Disagree	Don't know
(F)	25	66	9
(G)	18	52	30
(I)	13	81	6
(UK)	34	43	24
*	23	61	17

Statement: Tinned fruit and vegetables are just as good for you as fresh fruit and vegetables
Most children (60%) disagree with the statement that 'Tinned fruit and vegetables are just as good for you as fresh fruit and vegetables'. Just under one-quarter agree with it and around one-sixth do not know

%	Agree	Disagree	Don't know
(F)	25	68	8
(G)	20	50	30
(I)	15	78	7
(UK)	33	43	23
*	23	60	17

Statement: Fresh foods are safer than frozen or tinned foods
The majority (69%) of children believe that 'Fresh foods are safer then tinned or frozen foods.' Around one-seventh disagree with this statement and just under one-fifth don't know.

%	Agree	Disagree	Don't know
(F)	74	17	9
(G)	55	14	31
(I)	84	10	6
(UK)	63	14	24
*	69	14	18

* = four country average

Source: EUFIC

Chips are down for the canteen

As yet more children opt out of school meals, Nicholas Roe explores the great British packed lunch

Janet Stevenson is the sort of mother who makes other parents blush with guilt. Listen to her describe the packed lunch she gives her 15-year-old daughter Jade to take to school each day.

'Jade doesn't like school dinners,' says Janet, who lives in Brighton, East Sussex, 'so I usually make pasta or something she can eat cold. Pasta twirls with olive oil, onions, garlic, celery, carrots, tinned tomatoes . . . and at the end I put in some fresh basil and low-fat crème fraîche. I put it in those individual Marks & Spencer meal containers, which I save and then cover with cling-film . . . '

Hang on, she's not finished.

'Another one is diced chicken with whole grain mustard and crème fraîche; she likes that. And she takes yoghurts, fruit – apples, pears, plums, or satsumas – and mineral water, sometimes in a flask with ice and lemon, or I use one of those wine-cooling sleeves. She takes sandwiches, too, but I bake my own bread.'

When David Blunkett, the Education Secretary, gave school meals a C-minus for nutrition last week, he highlighted a trend embodied by Stevenson: most parents nowadays prefer their own food standards to those of the chip-encrusted school canteen.

The number of children taking school meals slumped from 64 per cent in 1979 to 43 per cent last year, with many families citing quality as a reason for change. But this itself poses a question: what do children find when they open their lunch box? And is it better or worse than the official alternative?

Judith Wood, a physiotherapist from Mattishall, near Norwich, says: 'I have five children and I do my husband's packed lunch as well, on a 7am production line.

'They all have their whims, but I have managed to work out a strategy where they get the same in principle, but I accommodate different tastes.'

The number of children taking school meals slumped from 64 per cent in 1979 to 43 per cent in 1996

Everyone has fruit, but the two youngest (Robyn, five, and Amelia, eight) get only grapes because, 'it's the only thing they will definitely eat. Bananas bruise and make other things smell and they don't like that; apples need more time to eat and they want to get back out in the playground. They come back half-eaten.'

Everyone gets some dairy produce: fromage frais, yoghurt or cheese triangles, as well as a chocolate bar. Then one or two rolls or rounds of sandwiches with sunflower spread and cheese or meat fillings.

The older children (William, 14, Gabrielle, 12, and Elizabeth, 10) sometimes have crisps or money to spend after school, though Judith admits: 'I live in horror of what they buy.'

Beware her early errors, which include sending a five-year-old to school with a bottle done up so tightly he couldn't undo it, so he didn't drink all day; and giving him so much food he missed playtime because teachers wouldn't let the poor chap out until it was all gone...

Jane Wheeler from Northfleet, Kent, makes packed lunches for three of her four children each night: Sarah-Jane, 14; Katie, 10, and Andrew, nine.

'It's like working for an army,' she says, ' but I do it because the children don't like school dinners

Food for thought

Serve Popular dishes, adjusting cooking to reduce fat	Avoid High fat/high sugar dishes, best served rarely
Beefburgers (grilled)	Cheeseburgers
Bolognese	Chips
Moussaka	Scotch eggs
Fishfingers (baked on racks)	Dumplings
Pasta	Toad in the hole
Risotto	Steak and kidney pudding
Jacket potatoes	Doughnuts
Banana custard	Ice cream
Frozen yoghurt	Lemon meringue pie
Currant buns	Shortbread

Source: DfEE guidelines to schools, February 1997

and Sarah-Jane is vegetarian and doesn't like the veggie choices at school. It's easier to make them a meal I know they will eat.'

Hers is a familiar choice – sandwich or roll with cheese or meat and sunflower spread; fruit juice; cereal bar; yogurt of fruit or both; low-fat crisps.

But although that sounds simple, it still wins the approval of Dr Michele Sadler, of the British Nutrition Foundation: 'Sandwiches are highly nutritious and a good balance of carbohydrate, protein and fat.'

'And if the children prefer white bread to brown, that isn't a problem either. The important thing is to eat starchy carbohydrate and all breads contain that.

'White bread also contains more calcium and iron because they are added – and calcium is one of the nutrients that some teenagers and children tend to go short of. If there's a yogurt in there as well, that's good.'

Many parents, including Christina Jackson, a child-minder from Northampton, fight a losing battle against crisps; last month her school-age children – Matthew, five, and Stephen, seven – devoured 24 packets between them, to go with sandwich, yoghurt, fruit and cake; but it was a trade-off, because this month they will have none.

It's a sensible deal, says Dr Sadler. 'Giving children what they like has to be part of the bargain. If they don't eat, they have no nutrient intake at all.'

Alina James from Travadlock, Cornwall, claims the triumph of never buying crisps or fizzy drinks for her three children (Thomas, 17; Leah, 14; Joshua, nine). A teacher herself, she is 'appalled' by school meals and insists on including raw vegetables in her children's packed lunch; typically red pepper, carrot and a chunk of cucumber. Home-made mayonnaise covers salad sandwiches – 'very fluffy and yellow'.

This still may not quite match Janet Stevenson's pasta sauce and home-baked bread. But if the bread rankles, parents can take heart. 'I have a bread-making machine,' admits Janet. 'It takes two minutes to put in the ingredients and turn on.'

© *Telegraph Group Limited, London 1997*

Junk food is taken off school menu

By Rachel Sylvester, Political Staff

The Education Secretary, David Blunkett, declared 'war on constant junk food' yesterday as he announced minimum standards for school meals.

Schools will be forced to provide food that meets compulsory nutritional specifications set by the Department for Education and Employment as part of the initiative, he said.

Nutrition 'hit squads' will tour the country and caterers serving 'chips with everything' will be warned that they could lose their contracts if they fail to reach the standards.

'School food has to be enjoyable and children need to have a choice,' said Mr Blunkett. 'We cannot ban chips – my own sons are fanatical about chips – but we should encourage more variety. This is a war on constant junk food.'

He told the Unison annual conference that the standard of school meals had fallen since the Government dropped national standards 17 years ago. He com-

plained that children ate chips on average three times a week.

The Education Secretary said he would 'consult widely' on how minimum standards could be introduced to make school meals less stodgy and more nutritious.

It was crucial for children's health that they had a 'balanced diet' at school. 'If you are hungry and you have a poor diet it is difficult to concentrate and to learn effectively.'

The Education Secretary said it was regrettable that the number of children eating school dinners had dropped from 64 per cent in 1979 to 43 per cent last year.

The midday meal was important for pupils, Mr Blunkett told the conference, because an estimated 30 per cent of children did not have a cooked meal in the evening at home.

Health campaigners welcomed

the move. Imogen Sharp, director of the National Health Forum, said the decision would reverse '17 years of nutritional chaos in the school canteen'.

Prof Gordon McVie, director general of the Cancer Research Campaign, said research showed that many mothers had given up trying to 'force feed a balanced diet at home'.

He added: 'We believe this is totally unacceptable as the cancers that can kill us are the ones in which nutrition plays the biggest part. There needs to be a shift back to healthy eating habits and school dinners is a good place to start.'

But the Government has some way to go to get its message across. Tucking into a baked potato with beans at Blatchington Mill school in Hove yesterday, Mr Blunkett asked pupils what they recommended from the canteen.

The response was unanimous: 'Chips.'

© *Telegraph Group Limited, London 1997*

No jacket potato required?

How has the changing nature of the working environment affected the nation's lunch-time eating habits? Ronnie Dungan finds out . . .

As companies continue to place more emphasis on profit and results, so the pressure on workers has increased to extend their output and increase their workload. This pressure has had an inevitable effect on the way in which workers view their own position in relation to their employers.

One way which this manifests itself is in the fact that the average worker is taking less time out in the middle of the day. People are having to work harder and the lunch hour is shrinking. During the 90s people are working longer hours and the lunch-break is getting squeezed until it becomes not so much a break as a mere fracture in the working day.

According to 1990 figures from Eurest, 73 per cent of workers took a regular lunch-break, now less than half of the UK's workers do – only 44 per cent. In fact, a large proportion of the working population, some 29 per cent, take no lunch-break at all. This figure has risen considerably over just the last seven years from seven per cent. Paradoxically, the average length of a lunch-break has increased from 32 to 33 minutes over the last two years.

Younger workers are the most likely to take a proper lunch-break – 68 per cent of 16-24-year-olds say they have one every day, whilst older workers (over 55) take the shortest breaks, with 70 per cent taking half an hour or less. In a regional context, workers in Lancashire take the longest. More than two-thirds (68 per cent) take over half an hour.

Trivial pursuits

Taking a lunch-break doesn't necessarily entail eating food either. Other than eating, reading is the most popular lunch-time pursuit for 32 per cent of the population. 'Relaxing', in whatever form it takes, is how 27 per cent of workers spend their break. Working is popular with 20 per cent. More people are now choosing (if it is, in fact, through choice) to work through the break than seven years ago when the figure was 17 per cent. Senior managers and professionals are the real workaholics, 27 per cent claim to skip lunch in order to keep working.

Japanese-type fitness kicks seem to have faded nicely out of fashion. Only five per cent of the workforce exercise through lunch now, compared with 13 per cent in 1990. Those of a more chauvinistic bent won't be surprised to learn that 24 per cent of women like to go shopping in their lunch-break – twice as many as men.

Lunch-time drinking seems to be less in fashion than it has been in recent years. Most workers (83 per cent) claim never to touch a drop during their lunch-break compared to a figure of 68 per cent seven years ago. The younger section of the workforce is most likely to partake in a lunch-time livener – seven per cent drink once a week or more during their lunch-break. Shucking their nationalistic stereotype, the Scots are the most abstemious, with 94 per cent of them claiming never to drink during lunch. It's in the Midlands and the North-East that workers enjoy their lunch-time drink the most, 20 per cent like a drop of the hard stuff.

Lunch money

But whether they're eating or just going for the liquid lunch, how much does the average worker spend every day? The national average spend on lunch at work is £1.45. This compares with an average spend of £1.20 in 1990 – an increase of 21 per cent, which is below the line of inflation. A third of the workforce still manages to keep its lunch-time spending down to less than £1. It's unlikely that many of these work in London, which is the most expensive, averaging £1.79. In Yorkshire, the average is the lowest at £1.09.

Of the myriad foreign foods now available on our High Streets, Italian is the lunch-time favourite, closely followed by Indian, Chinese,

Lunching habits

A Eurest survey of workers' eating habits found that nearly half of them take a packed lunch to work

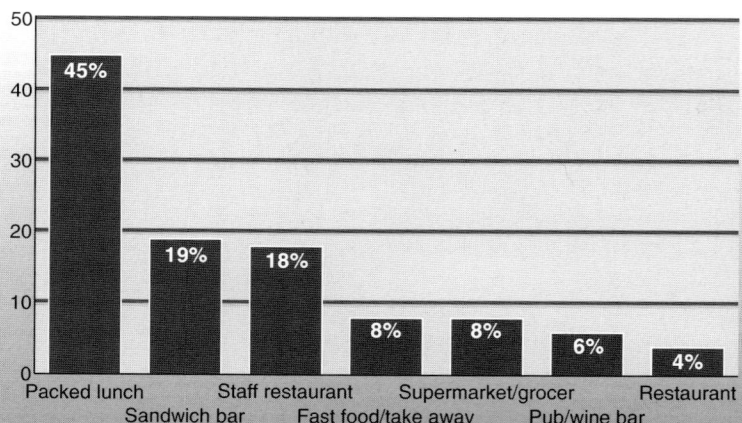

Category	Percentage
Packed lunch	45%
Sandwich bar	19%
Staff restaurant	18%
Fast food/take away	8%
Supermarket/grocer	8%
Pub/wine bar	6%
Restaurant	4%

Source: Eurest

American and Mexican. Workers in London have the more eclectic culinary taste, whereas in Ireland and Yorkshire workers expressed a dislike of foreign food.

Despite the vast array of low-fat, low-calorie, vitamin-enriched options available, health does not seem to be a serious consideration for the majority of workers. Over half (51 per cent) said they were not concerned about such things as sugar content, fat content and salt content.

That's not to say that the issue is simply being ignored and that there isn't a large body (large in numbers at least) that does count the calories and other things. Some 32 per cent take fat content into account, 24 per cent take note of the calorie count and 20 per cent watch the sugar levels.

The most health-conscious age group are the over-45s, particularly professionals and senior managers. Women, too, are concerned about calorie counting to an above average extent, some 37 per cent regarding it as an important consideration in their lunch-time choice. London and the Midlands holds the more diligent calorie counters as 32 per cent claimed to give the subject some consideration. In the North-East, calorie watching is presumably for soft southerners, as only 18 per cent monitored their intake.

With the recent furore over BSE and E-Coli, concerns about food hygiene and safety are also high on many people's agendas – 42 per cent in all. Concern was highest in London, where 54 per cent expressed their fears. Has this anything to do with the higher proliferation of kebab shops in the capital?

No future?

So what does the future hold for workers at lunch-time? Is the lunch-break set to become a thing of the past? Of the most popular suggestions from workers as to the future shape of their midday sojourn, 17 per cent believe there won't be a lunch-break at all and another 14 per cent believe it will be shorter by the year 2025. A quarter of 16-24-year-olds share the latter view.

A disconcerting 10 per cent actually believe that people will be taking their meals in a single pill in 28 years' time. They may, however, have watched far too many episodes of *The Jetsons*.

The growth of mobile working practices means that more people are now snacking and 'grazing' as they work, rather than setting aside time to eat a proper meal in the middle of the day. A lot more people eat lunch at their desk, which has created a demand for convenience and speciality foods.

If one in five workers are skipping their lunch-break altogether, it suggests that people are placing a higher value on work than on health and that the pressure placed on workers by employers is increasing.

© Direct Response
September, 1997

Grazing gets a grip and Britain becomes a nation of snackers

Put away the posh china – formal dining is dead. This was the conclusion of a new survey from business information company Key Note

The survey reveals that we are more bothered about what's on the telly than what's on our plate. TV dinners are here to stay and over two-thirds (65%) of over a thousand people questioned in the survey admitted that they regularly watch television while eating.

The trend for snacking has taken off and a significant 38% of those questioned said that they regularly ate snacks between meals. Young people were much more likely to 'graze' and 60% of 16-to-24-year-olds said they regularly ate between meals compared with 30% of 45 to 64-year-olds and just 22% of the over-65s.

Under half of those questioned (46%) said that they ate three proper meals a day, and a tiny 15% often ate a cooked breakfast.

But while we are not bothered about where and when we eat, most of us are careful about what we eat. 81% of those in the survey said that they often bought fresh fruit and vegetables, while over half of those questioned (57%) said that they regularly buy fresh meat and fish when doing the family shop at the supermarket. Under a third of those questioned (27%) said that they regularly eat convenience or micro-wave meals.

Quality is also important and over two-thirds (67%) claimed that they were prepared to pay more for premium products.

Young people's taste for the exotic means that traditional Sunday roasts and fish and chips will fall out of favour as this generation grows up. While 58% of 45 to 64-year-olds and a substantial 82% of the over-65s said that they preferred traditional British food to ethnic or European varieties, British fare was a hit with only 28% of 16 to 24-year-olds.

For further information or a copy of this report: Claire Williams Fannin, Templemere PR, 01483 283891, or Jonathan Thomas, Key Note, 0181 481 8767

- The Key Note UK Food Survey 97 was conducted by Gallup. A weighted sample of 1004 people was questioned. *©Key Note*

Britons 'the burger kings of Europe'

By Sean Poulter, Consumer Affairs Correspondent

Britons have munched their way to the top of the European fast food league – outspending their neighbours on hamburgers, pizzas and fried chicken.

Average spending on fast food has soared to £41 a year – equivalent to a McDonald's burger once every five days for every person.

In a study of the five major European economies, Britain's fast food bill was £2.42billion a year – way ahead of France, Germany, Spain and Italy.

The study did not consider other fast foods such as fish and chips, which would have added even greater weight to the argument that we prefer a quick takeaway than cooking our own healthy meals.

Government healthy eating messages appear to have been swamped by the enormous growth in the fast food business, especially among the big chains. Britons are also expanding – more than 35 per cent of us are considerably clinically overweight.

France ranked second with a fast food bill of £1.7billion, an average of £29 per head. Germany is third at £1.68billion, or £20 per head, followed by Spain at £427million, or £11 and Italy at £176million, or £3.

Hamburgers – which are relatively cheaper in Britain – dominate in all five countries. In France they accounted for 88 per cent of fast food sales. In Italy it was 86 per cent, and 85 per cent in Germany

In the UK and Spain, sales were more evenly divided between burgers and pizza, making pizza a more popular takeaway here than its Italian home territory.

Fried chicken sales took only a minor share in most countries and were almost non-existent in Italy.

'The UK is the fast food capital of Europe,' said a spokesman from Mintel, the research organisation that carried out the study.

'Fast food is benefiting from increasing trends towards snacking and convenience foods, coupled with the expansion of outlets led by McDonald's and Burger King.'

Nutritionist Dr Ann Ralph, of the Rowett Research Institute in Aberdeen, was dismayed at the findings.

'This is a worry, because the more people are buying fast food the less they are thinking about the balance of their diet,' she said.

'We seem to be losing the art of preparing a meal from real ingredients.'

© The Daily Mail
April ,1997

Italy succumbs to English taste

By Bruce Johnston in Rome

Italians now prefer drinking beer to wine and like an 'Anglo-Saxon breakfast', a survey claimed yesterday.

A report by Istat, Italy's national statistics institute, found that 61 per cent of men and 31 per cent of women choose beer over wine as their favourite alcoholic beverage.

The result contrasts with Italy's age-old tradition as a wine-drinking culture, but reflects a modern reality. In Rome it is now easier to get an imported draught or bottled beer in an establishment other than a restaurant than it is a glass of wine.

The recent mushrooming of English and Irish-style pubs has coincided with the disappearance of shops with bars selling wine and oil, and the osterie that once served up plonk from the tap and hearty food from a limited menu.

Researchers said wine consumption was still higher in Italy – the world's biggest wine producer – because it remained the standard mealtime drink. But this did not apply to pizzerias, where the beer flowed. The institute also said that seven out of 10 Italians admitted eating a breakfast which it called

Anglosassone. Although this appears to contain no bacon or eggs, nor black pudding – in the opinion of Italians the most abhorrent of British foods – it is so named because it is more than just the traditional biscuit and coffee.

An 'Anglo Saxon' breakfast appears to be a more copious version of the Italian variety and to include toast and marmalade, fruit, filled croissants and one or more cappuccinos, or tea.

© Telegraph Group Limited,
London 1997

Can't cook, won't cook, say women

Snacks and microwave meals replace kitchen skills

Women are losing the art of cooking, relying instead on frozen and ready-prepared food, a report said yesterday.

For a whole generation, a home-cooked meal means a packet that has been popped into the microwave, according to a survey of 1,000 women by *Bella* magazine.

The magazine found that the average woman spends around an hour a day in the kitchen, less if she is single. Most of that time is spent using the microwave and putting the kettle on for a cup of tea.

A fifth of mothers base family meals around processed foods such as burgers and chips.

Bella concludes that despite the popularity of female cooks such as Delia Smith and Sophie Grigson, women have lost 'the art and basic skills of cooking'.

One reason is that more are returning to work after having children than before, so they have less time to spend slaving over a hot stove. Another may be the end of traditional cookery lessons at school.

While some may welcome the end of women's servitude in the kitchen, there are fears that the trend is leading to unhealthy eating. Many women do not get regular meals and a growing number rely on snacks and treats.

The survey found that around one in seven females aged 15 to 24 have only one meal a day though two in three pensioners still have three square meals.

Two in three women admit to eating on the move, grabbing crisps and biscuits, though two-thirds also said they eat fruit between meals. In the 15-24 age group, the number who snack is up to four in five.

The survey questioned only women on eating habits, and not men, so it did not discover if husbands and fathers were contributing to the cooking. But the figures suggest that

---

A fifth of mothers base family meals around processed foods such as burgers and chips

---

when the women do not cook a proper meal, the family eat fast foods instead – which implies that men do not lend much of a hand.

Bella's editor-in-chief Jackie Highe said: 'Convenience foods have certainly been a boon to women who work. But it is a shame if cookery skills are not passed on to future generations and we all end up eating convenience food only.'

© The Daily Mail
November, 1997

Wales the target in campaign to promote healthier eating habits

By David Brindle

The Government yesterday set the first targets to make us eat more fruit and vegetables. People in Wales will be the guinea pigs of an experiment to try to increase consumption of fruit, vegetables and salad.

According to latest figures, 44 per cent of Welsh adults eat fresh fruit most days. Ministers want this to rise to 55 per cent within five years.

Similarly, 33 per cent of adults eat green vegetables or salad most days. The aim is to raise this to 40 per cent by 2002.

The targets were announced by Ron Davies, Welsh Secretary, in a speech to the Institute of Health Services Management (IHSM) conference in Cardiff. They come as the Government is searching for ways to close the widening health gap between rich and poor in Britain.

Sir Donald Acheson, former government chief medical officer, has been commissioned to report on possible strategies.

Mr Davies said incidence of long-term illness was 30 per cent higher in Wales than in England, cancer 10 per cent higher and NHS spending on each person 12 per cent higher.

The Government would address the effects of poverty, unemployment and poor environment but he urged the community to play a full part.

The 'fruit and veg' target is one of 20 set by Mr Davies which Welsh health authorities must have adopted in their plans by this time next year.

Some of the targets reflect previous ones in Wales, and in England under the Health of the Nation programme, for reducing incidence of conditions like cancers, heart disease and strokes, and for curbing drinking and smoking.

© The Guardian
June, 1997

Understanding obesity

Over a third of the population of many European countries is overweight according to health experts. Obesity, or overweight, is usually measured as the relationship between weight and height, called the 'Body Mass Index' (BMI). For an individual, the BMI is determined by dividing the person's weight in kilograms by the square of his or her height in metres (kg/m2). It is widely accepted that a BMI of between 25-30 indicates overweight and a BMI of greater than 30 reflects obesity. While these classifications are somewhat arbitrary, they form the basis of Table 1 which presents estimates of the percentage of the population which is overweight or obese in certain European countries.

In recent decades the prevalence of obesity has increased. For example, in 1980, obesity occurred in 8% of women and 6% of men in Great Britain; by 1993, this had increased to 16% of women and 13% of men).

The risks of obesity

Obesity and overweight are important health concerns since they are associated with an increased risk of mortality. Although premature mortality is hardly increased in the moderately overweight (BMI 25-30), the risk of disability, mainly from musculoskeletal and cardiovascular diseases, is significantly increased.

With obesity, however, it is well documented that the risk of mortality rises exponentially as BMI exceeds 30. Obesity is associated with several specific health risks including an increased incidence of hypertension (high blood pressure), increased non-insulin-dependent (maturity onset) diabetes and high blood levels of cholesterol and other lipids. Corrected for these risk factors, obesity in itself has been reported to

be an independent risk factor for heart disease.

The distribution of fat on a person's body is another important factor. Fat or adipose tissue tends to accumulate centrally around the waist (visceral fat), or more generally, including on the hips and limbs, where the fat is said to be 'peripheral' (or 'subcutaneous' i.e. under the skin). The result is to give people what is described as an apple or pear shape respectively (Ashwell et al., 1985). Individuals with central fat distribution have a higher waist-to-hip ratio (WHR) and are more likely to have disorders of fat metabolism and to develop diabetes and coronary heart disease than those who store fat peripherally with a lower WHR. This effect is independent of the Body Mass Index. The higher rate of coronary heart disease in males compared to females may be explained in part by the well-documented higher visceral fat levels in males.

Finally, the obese also carry an enhanced risk of certain types of cancers, such as those of the large intestine and the breast.

Balancing energy intake and energy use

Body weight maintenance over a period of time occurs when:
energy expenditure = energy intake

A person's daily energy intake – measured in kilocalories (kcal) – comes from the energy-containing nutrients in food i.e. fat, alcohol, carbohydrate and protein. These nutrients contribute 9, 7, 4 and 4 kcal respectively per gram of pure nutrient consumed.

Energy expenditure results from:
* The basal metabolic rate (BMR) is the energy required to maintain all body tissues including muscles

and organs and to regulate body temperature. It comprises approximately 60-70% of daily energy intake in sedentary individuals. Contrary to popular belief, BMR in obese people is higher than that found in lean individuals, which reflects the greater total amount of metabolically active tissue of obese persons.
* The thermic effect of food (TEF) or the energy which is lost as heat following a meal; this comprises up to 10% of daily energy intake.
* Physical activity which is the most variable component of energy expenditure in individuals. In sedentary people, exercise may constitute approximately 15% of total energy expenditure, whereas an individual who regularly exercises may expend up to 30% of his or her total daily energy output in this manner.

Energy expenditure =
BMR + TEF + Physical activity
60-75%   10%   15-30%

With these facts in mind, it is useful to look at the likely causes of the recent increase in overweight and obesity.

The effects of heredity

There is no doubt that obesity has a genetic component, although research has not been able to pin-point a specific metabolic defect in man. In the past few years, a number of potential differences in metabolism have been observed. For example, obese people seem to be less able to mobilise the fat stored in their adipose tissue.

It is clear, however, that genetic factors interact with environmental influences. Also genetic factors could not be responsible for the significant increase in obesity during the past 15 years since the gene pool of indigenous Europeans cannot possibly have changed significantly in such a short period. Clearly factors other than genetics have been at work.

The importance of food intake

In the past, many people who had weight problems claimed that they ate very little, but recently it has been found that most people, especially those who are obese, tend to under-report the amount they eat. In order to gain weight, a person's energy intake must exceed his or her energy expenditure.

In addition to the quantity of food consumed, the composition of food is an important risk factor for overweight and obesity.

The fat content of the diet

Many studies have demonstrated an association between body fatness and the fat content of the diet (Lissner and Heitmann, 1995). It is generally accepted that the high-fat content of the average European diet results in the consumption of more calories than people generally realise. The most important reason why high-fat diets have such a substantial impact on body weight is thought to be that they are very energy dense, having a high kcal (energy) value per gram of food. Therefore, it is easy to consume more than is necessary to meet energy needs. Cotton and Blundell (1993) refer to this as 'passive over-consumption'. Furthermore, researchers have reported that fat is less able to satisfy the appetite than is carbohydrate or protein. Finally, fat is the last metabolic fuel in the body to be oxidised for the provision of energy – the body preferring carbohydrate and protein as fuel. In short, it is easy to eat too much fat and not too easy to burn it off.

This has been demonstrated in an experimental study in which young, lean men were given three diets with either 20%, 50% or 60% of the calories contributed by fat. (Typical European diets contain about 40% calories from fat.) The men did not know which diet they were eating and the foods were cleverly disguised to avoid detection of the changes. When served the highest-fat diet, the men ate more

food than they needed without realising they were doing so. Conversely, when they ate the lowest-fat diet, they ate less than they needed.

When the experiment was conducted with the men maintaining a sedentary existence in a metabolic laboratory, they also ate too much when consuming the 40% fat calories diet.

However, if they were allowed to go about their normal daily activities then they ate less than they needed and went into negative energy balance on the 40% fat calories diet (Stubbs et al., 1995b). Thus, the typical European diet (containing about 40% calories from fat) can easily lead to overeating when people have a sedentary existence.

In a separate but similar experiment to the free-living experiment described above, all the foods were prepared so that the diets had 20, 40, or 60% calories from fat but had the same energy density (that is the same kcal/g). Here the men did not over-consume energy on any of the diets and succeeded in maintaining their energy balance. This indicates that the reason high-fat diets are more fattening is mainly because they have a high energy density.

Conversely, it has been readily demonstrated that low-fat, high-carbohydrate diets result in spontaneous weight loss even in subjects trying to increase food intake and maintain their body weight.

Sugar and other carbohydrates in the diet

It also seems that diets high in sugars are not more fattening than diets

high in carbohydrates in general. It is well known that there is an inverse relationship between energy from carbohydrate and energy from fat in the diet. This means that as the proportion of energy from fat in the diet decreases, the proportion of energy from carbohydrate increases. Furthermore, this relationship extends to suga. A recent, large epidemiological study showed that the lowest sugar consumers were two to four times more likely to be obese than were the highest sugar consumers.

Apart from the differing effects of carbohydrates and fat on appetite and food intake, several other factors differentiate the body's two main fuel sources:

- The body more readily metabolises carbohydrate (and protein and alcohol); dietary fat tends to go straight into fat stores. This idea is supported by the fact that human fat stores have the same composition as the fat eaten in the diet;
- The energy expended to store dietary fat as body fat requires only about 3% of the ingested calories; on the other hand, storing carbohydrate as body fat requires about 23% of the ingested calories. This is probably why the body does not usually convert carbohydrates into fat;
- Protein and carbohydrate have a greater thermic effect (TEF) than does fat, i.e. the higher the protein and carbohydrate content of a meal, the greater the amount of heat generated immediately after the meal. In practice, however, this effect is probably small.

Do taste preferences play a role?

A preference for a sweet taste is innate, with sensory preferences being higher in children and declining by adulthood. It is unlikely that the preference for fat is innate. The degree of overweight has been reported to be negatively correlated with preferences for sweet taste that is, people with a greater preference for sweet taste are less likely to be obese. On the other hand, several studies have reported a preference for high-fat foods by obese and formerly obese individuals. It has been suggested that the palatable combination of sugar and fat contributes a considerable proportion of energy to the diet, leading to over-eating and excess energy intake. Some studies have implied that the presence of sugar in combination with fat in certain foods leads to their greater consumption. An example of the relative importance of foods containing both sugar and fat in the adult diet is given by calculations using the Dietary & Nutritional Survey of British Adults, and the British National Food Survey adjusted to include foods consumed outside the home. These calculations demon-strate that only 14.2% of total sugar and 16% of total fat intake in the UK household diet is composed of foods in a sugar-fat combination.

Similar observations can be made for diets from a number of other European countries including Ireland and the Basque country.

The (beneficial) impact of physical activity

In spite of the increase in obesity over the last decades, there has actually been a parallel decline in food energy intake. According to UK National Food Survey data (corrected for energy intake outside the home), average per capita energy intake has decreased by 20% between 1970 and 1990.

Prentice and Jebb (1995) have stated that the data from the National Food Survey are corroborated by individual studies on food energy intake. They also concluded that the percentage of energy derived from fat, although high, has remained stable over those 20 years. Therefore, energy intake and fat intake cannot alone explain the apparent epidemic of obesity.

In affluent societies, few people are engaged in physically arduous jobs and most domestic situations are now characterised by labour-saving devices and central heating. None of these changes are readily quantifiable, but it is clear that most adults, whether lean or obese, are sedentary.

In the UK, two questionnaire surveys of activity found that 7 out of 10 men and 8 out of 10 women were physically inactive.

While children in Scotland in the 1980s were about the same weight and height as those in the 1930s, their energy intake was about one-sixth and one-quarter lower for boys and girls, respectively. This indicates that children use less energy – i.e., are less physically active – today than previously. Using modern techniques to measure energy expenditure, Livingstone reported that the amount of energy expended on physical activity declined with age in children and adolescents.

This indicated a worrying trend towards a sedentary lifestyle in adolescence.

It is notoriously difficult to measure physical activity accurately and few studies have attempted to do so. Thus Prentice and Jebb decided to look at data on television viewing and the use of cars as a proxy measure for inactivity. The average UK person now spends 26 hours per week watching TV compared with just 13 hours in the 1960s. Furthermore, TV viewing hours are greater in the lower socio-economic groups who also have a higher prevalence of obesity.

Physical activity does not offer a cure for obesity. Experts, however, do recommend a combination of diet and exercise because exercise helps reduce the loss of lean body mass which occurs during dieting thus helping maintain the body's basal metabolic rate. In particular, low intensity exercise such as brisk walking favours fat oxidation; as there is some evidence that fat oxidation is impaired in obese and post-obese people, this type of exercise may be particularly beneficial to such people.

Thus most experts now recommend that everyone should try to increase his or her daily amount of physical activity. Even the increase in physical activity associated with many small changes in habits such as using the stairs instead of the lift, walking to shops or moving around the office sometimes, instead of sitting all day, can make a significant difference to total energy expen-diture. For those already active, using a bicycle instead of the car or participating in some planned activities in leisure time can also help maintain a healthy body weight.

Apart from a possible role in preventing obesity, physical activity has other benefits. It affects metabo-lism and increases protective high density lipoprotein cholesterol. It improves the body's handling of dietary fat and enhances the body's ability to use glucose, thereby reducing the risk of diabetes in susceptible individuals. There is also increasing evidence that maintaining an adequate level of physical activity throughout life may help reduce the incidence of the so-called diseases of affluence such as coronary heart disease and certain cancers.

Conclusion

It is widely recognised that in spite of the fact that people have a high awareness about the importance of diet, obesity and overweight are on the increase. Adjusting to a diet with a lower proportion of fat than our typical European cuisine is still seen as an important part of a healthy lifestyle. However, the important new message is that keeping physically active through-out life will benefit everyone in terms of general health and should also help in the fight against the increasing problem of overweight and obesity.

• The above is an extract from *Understanding Obesity*, reference paper number 3 of a series from the European Food Information Council (EUFIC). For a full copy including references, please contact EUFIC at the address on page 41.

© *European Food Information Council (EUFIC)*

Slimmers' yoghurt claims to make stomach feel full

Chocolates, biscuits and sweets which claim to fill you up – thus helping you lose weight – could be just around the corner. Nutritionists however are more sceptical. Glenda Cooper, Consumer Affairs Correspondent, reports

Scientists yesterday introduced a yoghurt with an ingredient which they say could help you lose weight by fooling the body into thinking that the stomach is full.

A substance called Olibra, made from palm oil and oat oil, is said to trigger the chemical reaction in the small intestine which tells the brain that you have had enough to eat. The first yoghurts containing Olibra went on sale in Sweden yesterday and the developer of the product, Scotia Pharmaceuticals, said it hoped to follow suit in Britain.

But nutritionists said yesterday that more work was needed before it would be possible to say whether the yoghurt fulfilled expectation, and they asked Scotia Pharmaceuticals for more information on the product.

Olibra is made by taking palm oil and extracting ingredients which appear to activate sensors in the intestine which then release peptides into the blood. These in turn send messages to the brain that food is in the gut. Mixing palm oil with oat oil and water produces an emulsion which carries the Olibra swiftly into the small intestine.

Scotia says that the feeling of fullness lasts for three to six hours, reducing the temptation to snack between meals and lessening the desire for food. It claims that consumption of calories at the next meal is significantly reduced.

A trial carried out by the University of Ulster involving 29 men and women found that after eating the yoghurt calorie intake was reduced by 16 per cent. Fat intake was reduced by 22.5 per cent. The participants in the double blind trial ate breakfast and then lunch when they were given either a normal or an Olibra yoghurt.

At 5pm a buffet meal was served where participants could eat as much as they liked. The amount of food eaten by each volunteer was recorded by pre-weighing all foods and weighing the leftovers.

The yoghurt's makers insist this is not an appetite suppressant along the lines of controversial drugs because it uses ingredients which occur naturally in the diet, and activates natural reactions.

However, Tom Sanders, professor of human nutrition at King's College, London, and author of *You Don't Have To Diet*, said yesterday that more testing was needed: 'The company is trying to wheedle its way into selling a product . . . without testing for safety.

'The study is very short-term and it is not going to say whether it's going to work in the long term. It also takes quite a long time to get signals to the brain and most people wolf their food down in 20 minutes whereas the brain signals may take one or two hours.'

He added that even if the substance made you feel full, that was not necessarily the answer to controlling appetite. 'The reasons why we gain weight and overeat are really quite complex. The idea that obesity is due to not controlling hunger signals is not the whole story. Most people eat because of the social situation.'

Robert Dow, chief executive of Scotia, said yesterday that studies to see the long-term effects and any side effects would be carried out.

But Professor Sanders said the idea that the product was 'natural' and, therefore, safe was not acceptable: 'You need to have everything tested after BSE where things were natural but extremely nasty.'

© *The Independent*
January, 1998

HE'S ATTEMPTING TO LOSE WEIGHT

Ken Pyne

What the label doesn't tell you

Just how much do you know about the food you eat? Shoppers are being kept in the dark says the Food Commission's co-director, Sue Dibb. Her new book* lifts the lid on the secrets in our food

Hidden health hazards

Processed foods provide at least half of the unhealthy fat, sugar and salt in our diet, yet shoppers often have no way of knowing just how much is in their favourite foods.

Fats and sugars can come in many disguises and the food industry uses a whole range of cosmetic additives to make high fat, sugar or salty foods look more attractive or healthier than they really are.

And if you've ever tried to find out how much sugar is in a can of cola or fibre in a tin of beans, you'll know that such information is often absent from labels.

Nutrition labelling is voluntary – that means it's up to manufacturers to decide whether or not to tell us what we are getting. While many do provide some information, too many don't. Even when nutrition information is provided it can be for too few nutrients and hard to understand.

And nutrition claims that a food is 'low fat' or 'sugar-free' can be misleading. There's no law to define most claims.

Hidden additives

Most additives have to be listed on food labels, but not all. Flavourings don't have to be listed by name, and certain additives such as solvents, enzymes and others used as processing aids (but which may remain in the food) escape a listing altogether. And certain foods are exempt from having to declare any of their ingredients, including additives. These include:

- wine and alcoholic drinks
- some confectionery and chocolate
- cheese, butter, most milk and cream products
- unwrapped foods such as bread and cakes
- take-away foods
- eggs and farmed fish (may contain dyes fed to chickens and fish to enhance colour)
- citrus fruits and apples treated with preservatives on skins.

All additives, except flavours, have E numbers although it is increasingly difficult to find E numbers in ingredients lists. A growing number of manufacturers use only the chemical name of the additive rather than the E number in the hope that shoppers won't take as much notice.

Hidden genetically engineered ingredients

Until recently consumers were told that, with a few exceptions, there was no need for them to know whether the food they bought contained genetically modified ingredients. But faced with the introduction of genetically modified soya into a vast range of foods, the anger that this lack of labelling provoked, not only from shoppers but also from retailers and some food companies, has meant a rethink. Proposed new labelling laws may mean that some genetically modified soya will be labelled, but not all. Foods containing soya protein will need to be labelled but the many foods containing soya oil (often just listed as vegetable oil on ingredients lists), or additives such as lecithin which can be made from soya, will not be, nor will foods which are exempt from labelling such as take-away and restaurant food, or unwrapped bread which may contain soya flour.

And for other foods already on our supermarket shelves, such as cheese made with a genetically modified enzyme, or tomato paste made with genetically modified tomatoes, there is still no law that says they must be labelled – it is up to the manufacturers to decide whether or not to tell us.

low fat? solvents? soya?

enzymes? sugar-free

...I only wanted a tin of beans!

Hidden chemicals

When you are piling the fruit and veg into your shopping trolley do you worry that you're getting a hidden dose of unwanted pesticide residues along with all that goodness? The trouble is there's no way to tell. There's nothing on the label to inform us which added extras we might be getting. We can't see them and sometimes we can't even wash them off. We're advised to peel, top and tail our carrots and parsnips and peel apples for children because some have been found to contain higher than expected levels of the toxic insecticide, organophosphate. For the vast majority of produce, pesticide residues are within legal limits but an increasing number of people are choosing organic food, grown without the use of chemical pesticides and fertilisers, and, in the case of meat, without growth promoters or the routine use of drugs such as antibiotics, residues of which can sometimes turn up in conventionally produced meat.

Other chemicals contaminants can come from packaging. Clingfilm in particular has caused worries in the past with chemical plasticisers migrating into fatty foods. And there are newer concerns about chemicals which can mimic the female hormone oestrogen – so called gender-bender chemicals, including some pesticides, phthalates (used in plastics) which have been found in baby milks, and PCBs and dioxins, environmental pollutants from waste burning. The food we eat and the water we drink are the main way that many of these chemicals get into our bodies.

Hidden bugs

Cases of food poisoning have increased at an alarming rate and tougher, more virulent bugs are appearing. Modern large-scale farming has increased the spread of bacteria, but for consumers there is no way of knowing whether the foods they purchase are contaminated. Half of fresh and frozen chickens tested by the Consumers Association in 1996 contained either salmonella or campylobacter bacteria. Cross-contamination of cooked meat from raw meat was behind one of the

world's worst cases of E-coli food poisoning in Scotland in which 19 people died. In early 1997 the government was accused of watering down and then suppressing a damning report into hygiene practices in abattoirs which were likely to increase the risk of contamination and food poisoning.

The only safe way to treat foods, especially meat, eggs and foods made with them, is to assume that there might be a risk of contamination. That means ensuring good kitchen hygiene and cooking foods thoroughly, particularly if using a microwave oven. But the only way to tackle the problem effectively is at source, that is for hygiene practices throughout the food chain, from farm to supermarket, to be improved.

Hidden allergens

As many as 1 in 200 people are thought to have an allergy to peanuts and other nuts such as walnuts, almonds, brazil and hazelnuts. Many other common foods including milk, wheat and eggs can also trigger reactions in sensitive people. Yet for allergy sufferers it can often be difficult to know whether foods contain ingredients that might provoke a reaction. Campaigns for better labelling were started after the tragic death of one young girl who unwittingly ate peanuts in a lemon meringue bought in a department store restaurant.

Peanut oil (also known as groundnut oil) can turn up in quite unexpected places, even in children's lollies – but most likely it will only be listed as vegetable oil. Some companies now warn that foods may contain peanuts and the government has asked companies to improve

manufacturing practices to prevent cross-contamination. It is also urging the catering industry to be more aware of the problem.

Hidden animal products

Vegetarians and vegans need to be eagle-eyed shoppers when it comes to identifying foods and ingredients that may be of animal origin.

- animal fat can be found in biscuits, cakes and pastries;
- gelatin, made from animal bones and skin, is increasingly used in yoghurts, jellies, sweets and low-fat dairy products and spreads;
- whey, lactose, casein and caseinates are all derived from milk;
- constituents of eggs such as albumen, lecithin and emulsifier may be found in a range of products from chocolate to mayonnaise;
- many fruits, particularly apples and citrus fruits, are often waxed with shellac – an insect secretion – and beeswax;
- 'finings' for clarifying beers and wines may be derived from milk, eggs, fish or from mineral earths and seaweed, but the label won't tell you;
- cochineal is a 'natural' colouring made from crushed female Mexican cactus beetle;
- mono and di-glycerides of fatty acids are emulsifier additives which may be derived from animal sources.

Hidden beef ingredients

With concerns about BSE many people are wary of beef. But not all beef ingredients are easy to spot. Hidden beef ingredients or beef derivatives can turn up, often quite unexpectedly, in foods such as chicken gravy granules, chicken stew with dumplings, salami, stuffing mixes, ready meals, biscuits and Christmas pudding. Perhaps even more worrying for some parents is the hidden beef in baby foods. Many non-beef sounding savoury baby foods may contain 'meat extract', beef bouillon, beef stock or gelatin.

Hidden truth

Companies use a whole host of tricks of the trade to pull the wool over our eyes about what we are really buying. The big words on the front of the pack often tell only half the truth – that breakfast cereal may be a 'low-fat food' as claimed on the front, but you'll have to read the small print very carefully to know that it might also be high in sugar and salt and low in fibre, if it tells you at all. And other foods claiming to be 'free from' an ingredient or additives such as colours or preservatives, may not be as special as they sound.

There's no law yet for most foods to tell you just how much of the main ingredients you are getting. Fish fingers don't have to tell you how much fish they contain, and juice drinks can be as little as 5-10% juice.

Even the name of the product may not be all it seems. The main ingredient in tinned mince and onions, for example, can be mechanically separated chicken. Pictures of fruit on the label can give the impression the food contains much more than it does, and words like 'farmhouse', 'traditional', 'natural', premium' and 'wholesome' paint a rosy picture but without any further explanation are typically meaningless. And meat products like bacon, ham and sausages can come laden with added water – but the full amount is not always declared on the label.

Fancy packaging can disguise just how much – or how little – you are buying. And it's not just expensive products like luxury chocolates – plastic pots for yoghurt and desserts can come with a domed bottom to reduce the contents while giving the impression of a larger pot. And some manufacturers have even resorted to reducing the pack size and charging more. One juice manufacturer reduced the size of their pure orange juice carton from 1 litre to 750 ml, upped the price, while at the same time redesigning the smaller carton to make it taller and thinner so it actually looked larger!

Hidden exploitation

When we go shopping, most of us think little about the conditions of agricultural workers in developing countries who produce much of our food. Many can face poor pay, be denied trade union rights or suffer ill-health from exposure to dangerous pesticides.

In Costa Rica for example, around 100 people die and 10,000 are severely poisoned every year by the $55 million worth of pesticides drenched over Costa Rican vegetables and fruit destined for export. And in Kenya, workers on pineapple estates spray pesticides banned in developed countries. While minor food scares make front page news, this human misery, largely in developing countries, goes unreported.

Buying fairly traded produce ensures that workers not only receive a fair wage for their labours but that working conditions are safe. Campaigners for fair trade want to see supermarkets, with their huge buying power, use their muscle to ensure that all the food they sell has been produced safely and traded fairly.

What the Label Doesn't Tell You, by Sue Dibb, Thorsons, £6.99.

Your right to know

The Food Commission is campaigning for your right to know exactly what you are eating. Better labelling with clearer, standardised information will help, but labels alone are not enough. We also need regulations and standards to ensure that we feel confident that our food is safe and healthy.

We are promised a new Food Standards Agency which will put consumers' interests first. That should be a great step forward but we also need more openness, with less decisions being taken behind closed doors, more honesty and better communication from those charged with protecting our interests.

A 10-point charter. You have a right to . . .

- *The truth* – Nutrition labelling on foods to be mandatory, not just for the 'basic four' nutrients – energy, protein, fat and carbohydrate – but also for saturated and *trans* fat, sugar, sodium and fibre.
- *The whole truth* – Labelling of all genetically modified ingredients in food – not just some – to provide consumers with a genuine choice.
- *Nothing but the truth* – New rules on claims on foods and in advertisements to prevent misleading health and nutrition claims.
- *No exceptions* – There is no excuse to allow certain products – e.g. chocolate, baby food, alcoholic drinks – to evade the labelling regulations. Drinks should show the units of alcohol they contain.
- *A sustainable future* – More food to be grown with fewer chemical pesticides and fertilisers, and greater support for farmers to convert to organic farming. Labelling of produce treated with chemicals after harvest.
- *Reduced pollution* – Action to reduce contamination of food and drinking water with 'gender-bender' chemicals.
- *Improved inspection* – Strengthening food monitoring and enforcement to improve hygiene and safety practices throughout the food chain.
- *Transparent processes* – All government advisory committees and working parties to publish their minutes and a full public disclosure of members' interests or links to the food industry.
- *Open access* – Freedom of information legislation to allow data on food additives and pesticides to be available for independent scrutiny, before decisions are made to permit their use.
- *Consumer education* – To make the best use of label information there needs to be a national initiative to improve shoppers' knowledge.

Shoppers 'confused' by nutrition labels on food

By Chris Mihill, Medical Correspondent

Shoppers are confused by the nutritional labels on food, with many having little idea of whether sausages are healthier than yoghurt, a survey for the British Heart Foundation has found.

Nutritional tables on the back of food packaging were only consulted by one per cent of shoppers, and the tables seemed to conflict with claims on the front of packages, such as 'low fat' or 'high fibre'.

The research, by NOP, involved 1,000 shoppers. Over 50 per cent thought that a 'low fat' sausage had less fat in it than ordinary yoghurt, although it contains 80 times more.

At least 40 per cent of the survey did not know that the term sodium, listed in the nutritional tables, referred to salt.

The lack of familiarity with labels was further highlighted with the finding that 85 per cent of people thought that roasted peanuts contained more salt then cornflakes, when the reverse was true.

The researchers found that most people did not know how to interpret the amounts of nutrients listed in tables. Many did not know how much

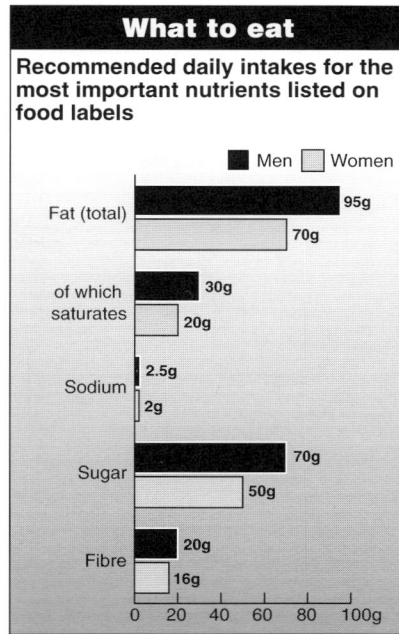

What to eat

Recommended daily intakes for the most important nutrients listed on food labels

■ Men □ Women

Nutrient	Men	Women
Fat (total)	95g	70g
of which saturates	30g	20g
Sodium	2.5g	2g
Sugar	70g	50g
Fibre	20g	16g

0 20 40 60 80 100g

fat should be consumed a day; only 35 per cent knew that men should eat less than 95g of fat a day.

A second piece of research, carried out by the British Heart Foundation's health promotion research group at Oxford, looked at the actions of people when they were specifically asked to shop for healthy food.

They were twice as likely to use the claims on the front of food packets as the nutritional labels on the back.

Even where nutritional tables were used, shoppers made unhealthy choices in 30 per cent of cases, either because they did not understand the information or were misled by the claims on the front of packages.

Overall, 70 per cent said they needed help in interpreting the nutritional labels, with 20 per cent saying the print was too small and 40 per cent that the information was confusing.

The foundation has produced a leaflet setting out what shoppers should look for in order to eat for a healthier heart.

It points out that claims such as 'low in fat' should be checked with care, as a bag of crisps that claims to contain 25 per cent less fat may still contain a lot of fat.

As a rule of thumb, the charity points out that 10g of sugar is a lot, whereas 2g is a little. In terms of fat, 20g is a lot, but 3g is a little, while 0.5g of sodium is a lot, but 0.1g a little.

© *The Guardian*
June, 1997

I NEVER SHOP WITHOUT HIM

BRITISH HEART FOUNDATION

Ken Pyne

Understanding food

In theory, the plethora of food products available today should facilitate the development of healthy personal and family diets. On the other hand, this presupposes that consumers know what is in these products and this can be difficult when they contain a variety of ingredients – for example, a pizza topped with cheese, salami and tomato; a ready-prepared stew with a combination of meat and vegetables; or a frozen mixture of several different vegetables. For this reason, increasingly informative labels on most processed foods can be very useful.

The first place to look for information on what a food contains is the ingredient list on the label. Ingredient lists give the components of the food product in descending order of importance by weight. So if vegetable fat or cheese, for example, appear high on the list, they are likely to be major ingredients, indicating a potentially high-fat food.

More information can be gained from 'nutrition information' now given on many labels. Energy (calories) and fat are the key things to look for.

Energy. From carbohydrates, fats, proteins and alcohol. On labels it is described as kJ (kilojoules) and kcal (kilocalories or Calories). A typical active man needs about 10,600 kJ (2500 kcal) per day; an active woman about 8,400 kJ (2000 kcal).

Protein. Essential for growth and repair of body tissues, protein is listed in grams.

Carbohydrates (of which sugars are often listed separately) are the major source of energy and include the important starches which should form the basis of the diet. While often thought of as 'fattening', carbohydrates contain just 4 kcal per gram, which is less than half the calories of fat. Carbohydrates are listed in grams.

Fat. Some dietary fat is needed to provide vitamins A, D and E and essential fatty acids. It is widely thought that many consumers in more affluent countries eat too much fat, putting them at risk of overweight or obesity. Saturated fat may also be listed. Current advice is that saturated fats should be only one-third of total fat intake (FAO/WHO). Fat is listed in grams.

Fibre. Found in vegetables, fruits and foods rich in starch, fibre adds bulk to the diet, helps satisfy hunger and may help protect against some diseases. It is often said that we eat too little fibre. Fibre is listed in grams.

Sodium. One of the components of common table salt, sodium enhances flavour and acts as a preservative in some foods. Some experts believe that those at risk of high blood pressure should reduce sodium intake. On the label, sodium is listed in grams.

Vitamins and minerals. There are many of these micronutrients that are necessary in very small quantities. Labels will list them in milligrams or micrograms and as a percentage of the RDA (Recommended Daily Allowance) of a particular vitamin or mineral.

Food laws require manufacturers to list nutrients per 100 grams of food or per 100 ml of liquid. If there is space on the label, manufacturers often add information such as nutrients per serving/portion. For example, on a breakfast cereal package, we may see information listed per 100 grams and per 30 gram serving. For some foods, the serving size may exceed 100 grams. The package may give information about the usual serving size of some products, while for others, e.g. a 330 ml can of a soft drink, the serving size is obvious.

People can learn a lot about the composition of foods from food labels on some fresh foods and most processed foods. This can be valuable for learning to predict what the composition of restaurant or take-away food might be. For example, a ready-made pizza in the supermarket could serve as a guide to the calories and nutrients that a restaurant pizza is likely to contain. So it can be worth investing time in thinking about what is on the label.

© *European Food Information Council (EUFIC)*

Foodborne illness

Its origins and how to avoid it

Safety is a priority at every stage of the food chain from farm to fork, and the foods available to European consumers are usually perfectly safe to eat. However, they may occasionally become contaminated to a level which spoils the food or could cause illness if eaten. This contamination can potentially occur at every point in the food chain, from harvest or slaughter to processing, storage, distribution, retailing, final preparation and serving.

Responsibility for ensuring that food remains in prime condition is shared by everyone involved in food production and handling. Strict legal controls are in place across Europe to ensure high levels of safety, hygiene and quality in commercial processing and food handling. Food manufacturers operate quality assurance systems to ensure safe food production, and few outbreaks of foodborne diseases originate at this stage in the food chain. The greatest risk to the consumer is through mishandling of food in establishments where food is served – such as restaurants, hotels, schools and hospitals – and in the home. The home in particular is an area where improved consumer awareness of food safety issues can reap real rewards in terms of risk reduction.

This article aims to answer two major questions. How does food become contaminated by microbes? And what can the consumer do to help ensure that the food he or she eats is safe?

Microbes in food . . .

Whether raw or processed, food is rarely sterile when it reaches the consumer. It usually contains bacteria or other microbes, most of which are harmless. Occasionally, it may also contain pathogenic microbes, which could potentially be a threat to food safety. The internal tissues of plants and animals have many defences to keep microbes at bay, with the result that healthy, freshly-harvested crop plants and fresh meat are usually sterile. However, changes after harvesting or slaughter, or during processing, may allow microbes to enter the food. These may originate from the crop plant or animal itself, from the environment, the factory environment (via soil or animals such as insects, birds and rodents) or from human sources.

Most food spoilage is due to microbial activity. Whilst spoilage does not necessarily make food unsafe to eat, it can make it unpalatable. Examples which pose little health risk are the moulds which can ruin the appearance of fruits and bread; yeast and lactic acid bacteria which can spoil sauces and beverages; and slime produced by microbes which can make chilled meat look unappetising.

Foodborne illness . . .

The World Health Organisation reports that, in spite of advances in modern technology and efforts to provide safe food, foodborne diseases remain a major public health concern both in developed and developing countries. In the UK, for example, foodborne illness affected one person in every 1000 in 1992, double the number of reported cases in 1987. In Sweden in 1992, there were more than 5000 reported cases of salmonellosis alone. While more accurate reporting may account for some of the recent increases in reported cases, it is clear that foodborne illness remains a problem.

There are two main categories of foodborne illness:

Infections

These result from eating food containing pathogenic microbes which then multiply in the body. There are two types:

- Infections where the microbes attack the intestines or other organs directly, causing symptoms such as nausea, vomiting, diarrhoea and fever. Here, there may be a gap of several days between eating the infected food and the appearance of symptoms, due to the time taken for the microbe to multiply. Examples are the infections caused by the bacteria Salmonella, Campylobacter and Listeria monocytogenes.
- Infections where symptoms such as diarrhoea result from poisons or toxins produced by the microbes as they grow in the intestine. Here, the time taken for symptoms to appear can vary from several hours to several days after eating infected food. An example is the illness caused by a toxic strain of the bacterium, Escherichia coli.

Intoxications

These result from eating food which contains toxins produced by microbes which do not need to grow in the human body to cause illness. Symptoms can begin only a few hours after eating contaminated food. An example is botulism, caused by a toxin from the bacterium Clostridium botulinum.

Special considerations . . .

Pregnant women, babies, young children and the elderly have particular needs which require special care in selecting, storing and preparing food. In general, people with reduced natural defences should particularly protect themselves against foodborne diseases. People taking certain kind of drugs, including antibiotics and chemotherapy treatment, are also more susceptible than normal to microbially-caused food disease.

Which foods pose the greatest risk?

Foods of animal origin are the primary source of many food poisoning microbes, such as Salmonella, Listeria, Campylobacter, E coli and L monocytogenes. These may occur on the live animal, and remain in the meat after slaughter. Without appropriate treatment to kill the microbes, or if conditions of hygiene or temperature control are poor, microbes may still occasionally be present in the final food product.

Foods which pose a relatively high risk of foodborne illness include:

Poultry, meat & eggs

The incidence of contamination is probably highest in poultry. Here, rapid growth in poultry production has resulted in a readily-available source of meat. However, there has been increased infection with food poisoning microbes in poultry, meat and eggs. Eggs can carry bacteria such as Salmonella enteritidis on their shells or within the egg. Salmonella infections are on the increase across Europe.

An important precaution in preventing foodborne illness from poultry and eggs is thorough cooking; the World Health Organisation recommends that raw egg should be viewed as a potentially hazardous ingredient which should not be used in foods which will receive no further heat treatment.

Red meats

These can also be contaminated with pathogenic microbes, probably to a lesser extent than poultry. The process of grinding meat to make mince and burgers may spread the microbes from one source into many products. As for poultry products, red meats should be thoroughly cooked before serving.

Dairy products

Raw milk can contain various pathogens from the dairy animal or its environment. Pasteurisation destroys all pathogens, and sterilisation ensures that the product is free from all microbes. Whilst pathogens are inactivated by many of the methods used to produce dairy products – including acidification and fermentation of milk – certain types may sometimes survive. Hard cheeses, yoghurt and butter are regarded as safe because of their acidity or lack of moisture, but mould-ripened soft cheeses can allow growth of Listeria monocytogenes.

Shellfish

As filter feeders which extract their diets from large volumes of water, shellfish can concentrate pathogens in their bodies. Inadequately heat-treated shellfish can cause a range of infections due to bacteria (such as Vibrio and Shigella), various parasites or viruses.

Herbs and spices

These frequently carry large numbers of bacteria such as Bacillus cereus, Clostridium perfringens and Salmonella.

How processing provides protection . . .

Commercially manufactured foods are designed to be safe. A number of common preservation methods are used to destroy microbes or stop them from growing, including:

- Heat treatments such as pasteurisation and sterilisation.
- Canning.
- Low temperature storage, for example, refrigeration and freezing.
- The addition of chemical preservatives, such as organic acids and nitrites.
- Natural antimicrobial products, such as nisin.
- Fermentation.
- Control of water content through drying, salting or smoking.
- Modification of the atmosphere in which the food is packed, for example, vacuum packing and gas packing.

Less widely-used preservation methods include high pressure and irradiation, together with novel technologies such as treatment with electricity or pulsed light treatment.

There remains a slight possibility that foods can become contaminated after processing. This can largely be avoided if everyone involved in handling food follows good hygienic practices.

Avoiding food contamination . . .

Food safety is a shared responsibility of everyone involved in the food chain from farm to fork. This includes primary producers, food companies, establishments which serve food, and consumers.

At the farm level, there are critical control points at every stage in animal rearing and crop agriculture where contamination of produce can be minimised by following good practices. After slaughtering, for example, inspections are carried out to separate diseased meat from healthy meat. However, even healthy animals can carry human pathogens and their meat can also become contaminated during slaughtering. These pathogens can be difficult to eradicate. Fresh fruits, vegetables and herbs can also become contaminated if they are fertilised with animal manure or come into contact with impure water. Crop plants may be treated to destroy pathogens – for example using biocidal washes – but such treatments are not always carried out.

At the food manufacturer level, the majority of companies have in-house quality assurance systems to ensure the safe production of food. Throughout the European Union, there are moves towards less prescriptive regulation and greater emphasis on industry responsibility. Measures currently used to help prevent contaminated food from reaching the consumer include:

- Using good quality raw materials from assured suppliers.
- Following Good Manufacturing Practices.
- Using management systems which allow the identification, monitoring and control of hazards during production, processing and sale of food.
- Providing training programmes for all food industry personnel.
- Carrying out research on pathogens and how best to control them.
- Exchanging information on food safety.

At the next level of the food chain from farm to fork, many foodborne diseases occur either as a result of mishandling in catering establishments or in the home.

A number of simple rules are recommended by the World Health Organisation to ensure the safe preparation of food:

- Avoid contact between raw and cooked foods, to reduce the risk of cross-contamination.
- Wash hands before preparing food and after handling raw foods, to minimise possible contamination.
- Cook food thoroughly in order to kill any microbes present. All parts of the food should reach a temperature of at least 70°C.
- Cool cooked foods as quickly as possible and then refrigerate. This slows down or stops microbial growth, which occurs best at 10-60°C.
- Reheat cooked foods thoroughly, to kill any microbes which may have developed during storage.
- Keep all kitchen surfaces clean to prevent cross-contamination.
- Protect foods from insects, rodents and other animals which may carry pathogenic microbes.

© *European Food Information Council (EUFIC)*

New food watchdog could cost shoppers an extra £14 a year

By Andrew Sparrow, Political Reporter

Shoppers are likely to pay the final bill for a powerful new Food Standards Agency, it was confirmed yesterday.

Ministers believe food producers and retailers are certain to pass on the estimated £200 million the health and safety watchdog will cost them.

That would amount to £14 a year on the average household food bill. Agriculture Minister Jack Cunningham insisted that shoppers would be getting a bargain.

He told MPs: 'The consequences of the BSE and E.coli outbreaks and salmonella poisoning have been colossal when compared with the costs of this agency.'

His view was endorsed by the Consumers' Association, which said

it hoped his plan would help end 'a decade of food scares and scandals'. Consumer groups believe the extra costs should be kept to a minimum so long as shops do not use the charges as an excuse for raising prices.

In a White Paper launched yesterday by Dr Cunningham, it is suggested that the FSA could raise millions by imposing a £100 levy on the country's 600,000 food premises, ranging from factories and caterers, restaurants and shops. But the Food and Drink Federation last night expressed fears that the charge would

become a 'food poll tax', which would hit the smallest shops hardest.

Dr Cunningham said the FSA's clear priority would be to protect public health.

Although the agency will also provide information about healthy eating, he stressed that it would not adopt a 'nannying' attitude.

'It will aim to help people who want to be healthier to choose a suitable diet,' said Dr Cunningham. 'It will not tell people what they must eat.'

Setting up the FSA will effectively shut down half the Ministry of Agriculture, Fisheries and Food which has long been seen as biased in favour of farmers and other producers.

The new agency, run by a commission of 12 independent experts and headed by a chief executive who is likely to become a household name, will take over MAFF's responsibilities for ensuring that food meets the highest standards. The agency will report back to the Department of Health, not MAFF.

Although the FSA is expected to cost more than £100 million a year to run, the bill to the food industry will be higher because local councils spend up to £150 million a year enforcing food standards and the Government wants producers and retailers to start paying for that as well.

The White Paper justifies the change by saying that the industry will benefit from improved public confidence.

Dr Cunningham said the idea of a £100 fee was still at a consultative stage. However, it is unlikely the Government will opt for a system that charges a tiny sandwich shop the same as a superstore, and the Minister even hinted that small businesses might be exempt.

The National Farmers' Union said it was 'strongly opposed' to any plans to charge food premises.

Beef ban breakthrough

The ban on British beef is set to be partly lifted after the first real breakthrough in Brussels since the trade blockade began.

The European Commission yesterday backed Britain's proposals to make an exception for meat from herds certified as being BSE-free for eight years.

This effectively limits any resumption in export trade to Northern Ireland's grass-fed prime cattle.

Ministers still face the battle to secure the agreement of at least seven other European allies for the scheme, which involves cattle whose certification is backed up by computer records.

But the National Farmers' Union yesterday welcomed the first 'glimmer of hope' for beleaguered farmers after 22 months of hardship.

Leader Sir David Naish said he was 'determined' that the UK should

press on with its demands for cattle born after August 1, 1996, when BSE controls began to come into effect, to be excluded from the ban.

England, Scotland and Wales are working towards a computerised record system – farmers were sent details this week – but may have to wait until the arrangements satisfy Brussels.

Yesterday's recommendation in Strasbourg came from the man who imposed the ban in the first place – EU Farm Commissioner Franz Fischler.

His decision is a welcome boost to Agriculture Minister Jack Cunningham. Angry farmers burned his effigy after he banned beef on the bone. But he insisted it was necessary partly to convince Europe that Britain was taking no chances in resolving the crisis over mad cow disease.

Dr Cunningham sounded a note of caution last night, saying: 'This marks a significant step forward in regaining international markets for British beef. This is only the first step however.

'We will be doing all we can to persuade the member states, as we have persuaded the Commission, that because British beef is safe there is no risk to public health in letting this scheme go ahead.'

© *The Daily Mail*
January, 1998

Food safety and your baby

Babies' immune systems are less developed than adults' putting them at greater risk of illness from harmful bacteria in foods. So, it's important to follow the tips below:

- Check that safety seals are intact when first opening food packaging.
- Read labels carefully and follow all instructions on storage, preparation and cooking times.
- Defrost frozen ingredients thoroughly in the fridge or microwave.
- Keep raw products separate from ready-to-eat foods.
- Cook foods thoroughly until piping hot, then allow to cool until comfortable to eat.
- After heating food or drinks in a microwave, stir them thoroughly and allow to stand for a couple of minutes to avoid hot or cold spots.
- For young babies, wash bottles in hot soapy water, rinse and sterilize using boiling water, a steam sterilizer or sterilizing solution used according to instructions on the label.
- When adding water to baby foods, milks and drinks always use cooled, boiled water.
- Keep kitchen surfaces extra clean.
- Teach young children to wash their hands before touching food, after using the toilet and after playing with pets. Make sure you lead by example by washing your hands too!!!
- Keep dirty nappies away from food and food preparation areas and always wash hands after handling dirty nappies.
- Store any extra food prepared (but not served) in a clean covered container in the coldest part of the fridge (between 0-5°C) and use within 48 hours. Never leave food in open cans.

These simple precautions will help protect your child from the dangers of food poisoning:
- Be a clean cook
- Control food temperatures
- Keep bacteria at bay
- Store foods safely.

© *Merton Environmental Health Department*

Danger labels are slapped on eggs, cheese, carrots, baby milk – and meat

In the last 20 years, one food scare has followed another, with panic reaching a peak over beef. Glenda Cooper, Consumer Affairs Correspondent, follows the trail of poison.

'We do warn people now that sadly most of the egg production in this country is infected with salmonella.' Thus the first big food scare of recent times was started in December 1988 by Edwina Currie, then a junior health minister.

It was the first of many. Ready-cooked poultry and soft cheeses were the next victims, with the listeria outbreak of 1989. The same year there was a botulism scare after one contaminated batch of hazelnut yoghurt was found.

In February 1993, high levels of patulin, a toxin occurring naturally in juice, were found in apple juice. And then in May 1995 the discovery of high organophosphate levels led to government advice to peel carrots before eating them.

Parents panicked in May 1996 when nine leading brands of baby milk were said to contain levels of phthalates, 'gender bender' chemicals. The European Commission later concluded that there was no danger to babies. Then 20 people died in an outbreak of E.Coli 0157. It began in November 1996, and was traced to a butcher's shop. The Government ordered an inquiry, led by Professor Hugh Pennington, which called for sweeping changes in food safety. This week, he said the Government had not acted fast enough on his recommendations. But the mother of all the scares is beef – specifically, the link between bovine spongiform encephalopathy (BSE) in cattle and new-variant Creutzfeldt-Jakob disease (v-CJD) in humans.

It was 1985 when scientists from the Ministry of Agriculture, Fisheries and Food (MAFF) began to investigate symptoms of a new disease – later named BSE – in cattle on dairy farms. In 1988, the Government set up a working party under Sir Richard Southwood to consider the significance of the BSE epidemic. Compulsory slaughter and incineration or burial of cattle showing symptoms followed. More than 170,000 cattle were diagnosed between 1987 and 1997, but estimates showed that up to one million with the disease but showing no symptom were used in human and animal food.

The Government insisted that there was no danger to the public. Many scientists, meanwhile, quietly gave up beef.

In May 1990, John Gummer, then the Minister of Agriculture, infamously fed his daughter Cordelia a hamburger to prove its safety. In 1993, Kenneth Calman, then chief medical officer, issued a statement to calm fears. Douglas Hogg, Mr Gummer's replacement, continued to emphasise that British beef was safe to eat.

But in 1994 a handful of teenagers fell ill apparently of CJD, which usually affects only those over 60. The Government denied a link to BSE.

On 20 March 1996, Stephen Dorrell, then Secretary of State for Health, announced that the most likely cause of v-CJD was exposure to BSE. A Europe-wide ban on British beef followed, killing a £500m industry almost overnight, and the Government banned sales of specified material from cattle, sheep and goats. By August 1997 more than 1.7 million cattle had been slaughtered. Most recently, the Minister of Agriculture, Jack Cunningham, banned the sale of beef on the bone.

To date, 23 people have died of v-CJD, but scientists say it is too early to know whether thousands more will follow.

© The Independent
January, 1998

E numbers

This pan-European study concerning children's views on food and nutrition was commissioned to obtain a basic understanding of the perceptions of European children on food- and drink-related issues. Research was focused on four countries, France, Germany, Italy and the UK. The majority (58%) of children do not know whether 'Foods containing E numbers are bad for you', whilst 29% agree with the statement and 13% disagree with it.

	Agree (%)	Disagree (%)	Don't know (%)
	19	17	65
	22	8	69
	44	9	47
	29	18	52
Average	29	13	58

Source: EUFIC

Eating outside the home

The risks involved. The consumer must follow basic safety precautions when eating in public places

Eating out

One step in the food chain from farm to fork which deserves special attention is the preparation of meals outside the home – restaurants, hospitals, nursing homes, child care establishments, schools, canteens, airplanes, wedding parties, business conventions, etc.

Considering the vast number of meals consumed in such public settings, professional chefs and caterers do an outstanding job of protecting the public against foodborne disease. However, while the incidence of problems in public places is very low, when something does go wrong, it tends to attract wide public attention. Since they usually effect a large number of people, such incidents are more often than not investigated by an official inquiry. In contrast, individual incidents of disease in the home are rarely reported at all.

As a result of investigations into institutional incidents, efforts have been made to inform and educate professional food handlers as well as the consumer in an attempt to avoid similar problems in the future. Such investigations have led to strict industrial hygiene procedures and, specifically, to the adoption of a system known as Hazard Analysis Critical Control Points (HACCP) which is used to ensure quality in food processing plants.

Some special procedures followed in public settings are:

- Where large numbers of meals are prepared in advance and/or far from the point of eventual consumption – schools or nursing homes or on trains and airplanes – facilities for hot holding or cooling must be carefully controlled to avoid problems resulting in people becoming ill from eating contaminated food.
- When large groups of people are served meals prepared on lines that are not designed for such quantities, catering personnel should be particularly careful to avoid contamination.
- At public events where food should look artistic as well as taste good, food service professionals face additional challenges. Raw decorations (parsley, shrimps, etc.) may result in food being handled more than is usual to acquire the desired effect. As a consequence, undesirable micro-organisms may enter an otherwise safe food dish.
- Lastly, while the food processing industry produces a relatively limited number of products using relatively standardised methods, food service establishments change recipes daily and often prepare several different dishes at the same time and in the same facility. Under these circumstances, professional food handlers should take particular precautions to avoid cross-contamination among raw materials or between raw food and finished products.

In spite of all these efforts, there are occasional incidents of foodborne disease originating in public places. Therefore, the individual consumer also has an important role to play in ensuring his or her own safety.

A consumer can lower risk of infection by only eating food which is piping hot, refusing food which has been left standing for too long, and avoiding food which should be cold but which is in fact lukewarm.

As in the home, food that looks or smells odd should be avoided.

Because hygiene in the kitchen is so important, the consumer should look for recognized certificates of good hygiene in public places.

A consumer who suffers from food allergies should be particularly cautious in public places, even avoiding certain food unless assured that there is no risk of contamination.

All in all, paying attention to the obvious signs of food quality is especially important in order to protect individuals with a higher risk of foodborne disease. © EUFIC

Causes of foodborne disease in service establishments

% of outbreaks where location was identified

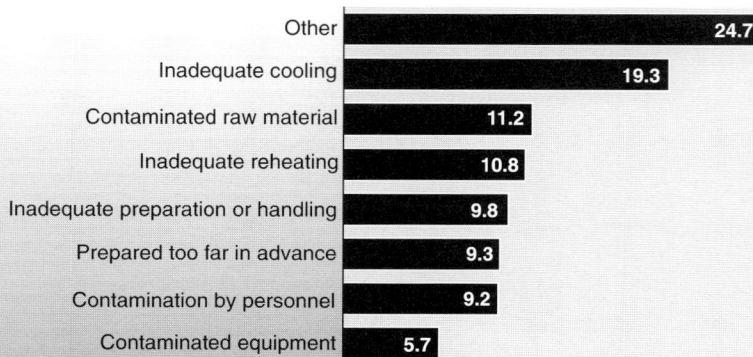

Location	%
Other	24.7
Inadequate cooling	19.3
Contaminated raw material	11.2
Inadequate reheating	10.8
Inadequate preparation or handling	9.8
Prepared too far in advance	9.3
Contamination by personnel	9.2
Contaminated equipment	5.7

Source: WHO Europe Surveillance Report 1990-1992

Food poisoning

The facts

In many cases the general public are aware of the importance of good food hygiene but do not always put this into practice in their own homes.

In a recent survey 7% of adults questioned believed they had suffered from food poisoning in the last year, an increase of 3% on the previous year.

In England and Wales the number of notified cases of food poisoning reached 82,587 during 1994 compared to 6,111 in 1949 when food poisoning notifications first began. Don't become a statistic!

To avoid food poisoning within the home take the following simple precautions:

1. Raw and cooked food should be prepared and stored separately.

This will prevent raw juices containing food poisoning bacteria from coming into contact with cooked food.

It is recommended that separate work surfaces and equipment are used for preparing raw and cooked foods. If this is not possible then work surfaces and items of equipment should be thoroughly washed between handling raw and cooked food.

Kitchen worktops and equipment should be kept clean.

Raw meat and fish should be stored at the bottom of the refrigerator, below cooked foods.

2. Food should be thoroughly cooked so as to kill any bacteria

Always follow manufacturers' instructions for preparation and cooking of food.

If food is to be reheated it should be thoroughly heated until piping hot. Inadequate reheating may not kill food poisoning bacteria which may be present in food and illness may occur.

3. Food should be stored at the correct temperature. This will prevent the growth of food poisoning bacteria within food.

Chilled and frozen foods should be taken home as quickly as possible and then placed in the fridge or freezer. On particularly hot days it is recommended that cool bags or ice boxes are used for taking food home.

Manufacturers' instructions on correct storage of food should be followed at all times.

It is important that household fridges and freezers run at the correct temperature. The coldest part of your fridge should be kept at five degrees celsius (5°C) or below whilst freezers should operate at minus eighteen degrees celsius (-18°C). It is recommended that fridge and freezer thermometers are provided to ensure that units are working correctly.

4. Hot foods should be kept hot and cold foods should be kept cold to prevent bacteria from multiplying.

Once cooked it is recommended that food is consumed immediately. If hot food is not to be eaten immediately it should either be cooled (but not for too long!) and then placed within the fridge until needed or it should be held at a hot temperature of 63°C and above. Food should not be left standing around at a warm temperature.

5. Hands should be washed thoroughly.

Hands should be washed, using hot water and soap, after going to the toilet, after handling pets and in between handling raw and cooked foods. Clean hand-drying towels should be used.

6. 'Use-by' dates should be checked before food is used.

This will ensure that food is only used within the manufacturers' recommended period.

7. Do not eat food containing uncooked eggs.

Raw eggs often contain the bacteria Salmonella. This bacteria is destroyed if eggs are thoroughly cooked but may cause illness if undercooked or raw eggs are eaten. Eggs should be kept in the fridge.

8. Pets should be kept away from food, equipment and worktops.

© Merton

Food poisoning

This pan-European study concerning children's views on food and nutrition was commssioned to obtain a basic understanding of the perceptions of European children on food- and drink-related issues. Research was focused on four countries, France, Germany, Italy and the UK.

Question: How do you get food poisoning?

%		Bacteria	Pesticides	Additives	Eating too much	Don't know
	(F)	43	20	13	30	21
	(G)	80	28	12	12	9
	(I)	56	50	28	24	4
	(UK)	65	15	2	2	17
Average		61	28	14	17	13

Source: EUFIC

Beef: a crisis out of control

Marrow link to BSE brings new bans

By Ewen MacAskill, James Meikle and Stephen Bates

A devastating new disclosure about BSE in British beef yesterday killed off any lingering hope of an early lifting of the European embargo and dealt a blow to consumer confidence and the domestic market's fragile recovery.

The Government, after evidence that BSE had been detected in bone marrow, was forced to extend its ban to cover T-bone steaks, ribsteaks and oxtails, which account for about 5 per cent of the British market.

Butchers will have to debone the meat before selling it on to restaurants and shops. Foreign meat will receive the same treatment. Tesco and Sainsbury's last night withdrew beef-on-the-bone products from shelves. The Consumers' Association advised people who wanted to avoid all risks to avoid eating meat altogether and said consumers may have been given false assurances in the past.

The Government was caught on the hop after a leak forced it to rush its announcement. There was initial confusion over the range of products affected by the ban. Makers of Bovril, for example, insisted its beef ingredients came from abroad and were unlikely to be affected.

Fresh oxtail is banned but makers of oxtail soup said most was produced with meat that had been deboned. The Government confirmed it did not see any significant problem for canned oxtail soup.

A spokeswoman for the Ministry of Agriculture, Fisheries and Food said commercial stocks and cubes could be affected by the latest regulations. 'Bones would not be allowed to be used to create soup or stock,' she said.

The Government also insisted that gelatin – a beef product made from beef bones used in a range of foods, from sweets to biscuits and Oxo cubes – could only be used if it had been 'satisfactorily demonstrated it was BSE free'.

Yesterday's disclosure came only two weeks before the European Union was due to discuss a partial lifting of its export ban. Northern Ireland and Scotland, which keep better records of their herds, had been in line for possible exemption from the boycott. But the first signals from Europe yesterday indicated that was now unlikely.

Tony Blair told the Commons: 'I do understand the plight the beef farmers face at the moment and how great a blow this must be to them, but when we receive scientific advice and the chief medical officer gives us recommendations we feel that we are under an obligation to follow them. We will, however, do everything we can to mitigate the problems.' He gave no indication whether beef farmers would receive increased compensation.

The new ban will hit beef prices and force many farmers out of business. The National Farmers' Union described it as 'a body blow'.

The domestic market had staged a partial recovery since the link with BSE was established in March last year. The Meat and Livestock Commission put consumption in Britain in 1995 at 901,000 tonnes before slumping to 739,000 tonnes last year, and, until yesterday, had predicted a partial recovery this year to 839,000 tonnes. The commission said Oxo and other beef stocks were not affected because they were made from Australian beef. The ban will not apply to dog food.

The Minister of Agriculture, Jack Cunningham, insisted in a Commons statement that there was only a 'very small risk' from beef on the bone. 'British beef is as safe as any in Europe and safer than most,' he said.

Labour in opposition blamed Conservative hostility towards Europe for failure to get the export ban lifted. Michael Jack, the shadow agriculture spokesman, said yesterday Mr Cunningham had been long on rhetoric and short on action.

In Brussels, although the European Commission response to Mr Cunningham's announcement was that it would have no effect on the speed of lifting the beef ban, there was private concern at the likely political response from member states to the news that a new risk had been discovered in British beef.

A Commission official said: 'Bone and bone marrow have always been considered to be low-risk materials. So long as Britain is just acting in connection with its own meat for its own consumers we would see no reason to intervene.'

BSE remains a highly sensitive issue on the Continent, where beef consumption is still down and where Belgium and Luxembourg have both recently reported their first recorded cases of the disease in local cattle. Other member states remain suspicious of the effectiveness of British attempts to tackle the crisis.

The ban on cuts of beef on the bone was one of three options from the Government's Spongiform Encephalopathy Advisory Committee. The others were simply to publish the information or to debone cattle aged between 24 and 30 months in specially licensed plants. Professor

John Pattison, the committee's chairman, said: 'We are dealing with something that is very small indeed … we are getting to the stage where BSE in cattle is no longer an issue for human health with all the restrictions we have in place.'

The new ban follows evidence from government-sponsored tests that infection from sensory branches of the nerves near the spinal cord of infected cattle was transferred to mice. The spinal cord is already removed from food for human consumption but these other tissues, dorsal root ganglia, would be left with the bone when meat was cut off the spine. Bone marrow was also infected in one test.

The new risk

1 T-bone steaks: 3,000 tonnes sold in Britain for domestic consumption, worth £40 million a year
2 Rib steaks: 25,000 tonnes, worth £110 million
3 Oxtails: less than 1,000 tonnes

Other products
Oxtail soup?
No cooking at home on the bone. Tinned soup will depend on how it is made.

Beef stocks and drinks?
Safe if made with beef products from a country where a BSE case has never been proved, e.g. New Zealand.

Other products with gelatin such as biscuits, sweets and fizzy drinks?
Thought to be safe, although confusion over full range affected.

The new research shows a possible contamination of 'dorsal root ganglions', peanut-sized nerve packets which are encased within the vertebral column. When meat is deboned, the ganglions are removed, but experts fear when meat is sold on the bone, there is a chance they may remain.

© *The Guardian*
December, 1997

Doctors warn that meat is dangerous

By Kathy Marks

The British Medical Association believes that too few people are aware of the dangers surrounding food preparation, and of how to avoid them. Last year, a record 1 million people were struck down by food poisoning, 200 of them fatally.

In a submission today to an inquiry into food safety by the House of Commons Agriculture Committee, the BMA says that all raw meat should be handled as if it were contaminated.

It adds: 'The current state of food safety in the UK is such that all raw meat should be assumed to be contaminated with pathogenic organisms.

'The only safe approach for the food industry and the general public is to treat all raw meat as infected and adopt "universal precautions" in handling and cooking raw meat.'

This week, the Government is

expected to publish a White Paper, recommending the establishment of a new Food Standards Agency, in the wake of the scares surrounding BSE, salmonella and E.Coli.

The BMA calls for re-education of the public, more information on food handling and detailed cooking instructions on labels, particularly for microwave cooking.

It says it is vital that food be chilled to below 5°C to prevent infection, but that too few people – food industry workers, as well as the general public – ensure that their fridges are working properly.

Doctors also want better nutritional information on packets in relation to sugar, salt, fat and fibre contents.

They say that descriptions such as 'low fat' or 'healthy' should be subject to stricter regulation, and that labels ought to make clear whether food contains pesticide residues or genetically modified organisms. However, they do not state whether they believe nutrition should come under the remit of the new agency.

Sandy Macara, chairman of the BMA, said yesterday: 'The consumer needs help in untangling the messages about what is good for us and what the real risks are.'

The BMA wants to see a new 'precautionary' approach to the introduction of new practices and ingredients into the food chain.

Dr Macara said it would be a 'dreadful calamity' if routine use of antibiotics in animal husbandry led to human resistance to their effects.

© *The Independent*
January, 1998

Food industry backlash over 'lunatic' measures

The Government faces a backlash from the food industry with accusations that the new safety measures are 'lunacy followed by panic'.

Farmers and butchers are preparing for the worst, warning that the beef industry may have to have millions of pounds of meat destroyed.

Supermarkets are stripping their shelves of premium steaks while restaurateurs hastily rewrite menus.

Around 11,000 family butchers throughout the UK will have to change their traditional ways of working.

They fear that if the ban is imposed in a draconian fashion their practice of maintaining quality and flavour by leaving it to the last minute before cutting meat from bone will have to end. Sides of beef may be boned out in central plants under official supervision and the flesh vacuum-packed and sent out around the country.

Specialist butchers whose trade is predominantly the supply of bone-in meat to restaurants and hotels fear they could be put out of business altogether.

The industry is already reeling from the effects of the strong pound and with inspection costs now likely to soar the consumer is bound to suffer.

David Lidgate of 150-year-old butchers C. Lidgate in Holland Park, West London, said: 'It's not the cows that are mad, it's the Government. It's lunacy followed by panic.' John Fuller, of the National Federation of Meat and Food Traders, said: 'The real issue is what this does for consumer confidence. The fact that these traditional cuts will change will disappoint people.

'The number of CJD cases is extremely small. BSE was first identified in 1986, but where is the evidence of the great epidemic that the doom merchants have been peddling?'

Elizabeth Sunley, assistant director of the British Meat Manufacturers' Association, whose members make pies, burgers, sausages and ready meals, said the measure would have a knock-on effect for all beef products.

'It's not the cows that are mad, it's the Government. It's lunacy followed by panic'

Tesco, Sainsbury, Safeway and Marks & Spencer all announced that they would no longer sell beef on the bone with immediate effect. A Marks & Spencer spokesman said rib of beef sold at 60 stores had been removed from shelves.

Tesco offered refunds to customers who had recently bought beef on the bone from stores and also cleared shelves of nine types of rib and boiling beef products.

Safeway said rib of beef would now be sold off the bone in its 444 stores at the same price per pounds – effectively giving customers more meat for their money. The Food & Drink Federation, which represents retailers and manufacturers, said companies selling beef stock cubes and soups in the UK had eliminated any risk by producing them in BSE-free countries – mainly Sweden and Australia.

The Pet Food Manufacturers' Association admitted that member companies might have to change their manufacturing procedures. A spokesman said: 'We would like to reassure pet owners that they should continue to feed their pets prepared pet food with confidence.'

However he said that it would take some time to be certain that the latest findings from SEAC would not require changes to manufacturing.

Many restaurants also reacted swiftly. The large Beefeater chain has removed T-bone steaks from its 300 restaurants.

Chef Michel Roux, of London's Le Gavroche restaurant, complained: 'If it's come to banning the sale of beef on the bone they may as well go the whole way and ban beef. The Government should be more supportive of the beef market.'

Carl Smith, manager of the renowned Guinea Grill in Mayfair, said: 'Steak is what our reputation is founded on. I'm not a scientist but this does seem to be ultra-cautious.

'I think with all the precautions in force already. Britain must have the safest beef on earth.'

© *The Daily Mail*
December, 1997

TV food executive hits out over scares

Cookery programmes producer claims public is left unduly worried in 'climate of ignorance'. By Kamal Ahmed, Media Correspondent

A senior BBC food programme producer yesterday attacked campaigning groups for causing 'unholy confusion' about food safety.

Paul Bazalgette, executive producer of the *Food and Drink* programme and the man behind television hits *Can't Cook, Won't Cook* and *Ready, Steady, Cook*, said people had been subjected to a 'roller coaster of food scares'.

He said the Government, scientists and consumer groups were responsible for worrying the public about products as diverse as coffee, eggs, red meat and fresh fruit.

The attack is the latest in a series of broadsides which reveals a new scepticism among broadcasters about the claims of pressure groups and environmentalists.

Both Channel 4 and the BBC have recently run series undermining green claims about the stripping of the rain forests, pollution and global warming, and saying that predictions of doom in the 1980s have not been borne out by events in the 1990s.

The Advertising Standards Authority also recently criticised the Vegetarian Society for making a direct link between eating red meat and cancer in one of its newspaper adverts.

'We live in a climate of ignorance about food and ignorance breeds fear,' said Mr Bazalgette in the latest issue of *Feedback*, the magazine of the Food and Drink Federation.

'We have to be aware of the dead rat syndrome. There's always a laboratory somewhere feeding some poor rodent vast quantities of material and then they are surprised when they die of cancer. Food is blamed far too much.'

He said that in the scare over the pesticide Alar, which was used on apples, the pesticide was found to cause cancer in rats when given in large quantities. Stars, including Meryl Streep and Pamela Anderson, launched a campaign against the pesticide, which was eventually withdrawn. It was later pointed out that mushrooms contained exactly the same carcinogen in far greater quantities.

'Red meat is a fantastic source of healthy nutrients and a fantastic source of pleasure'

Mr Bazalgette said that stories that microwaved milk might cause brain damage in babies was based on a research project which boiled infant formula until it was brown and solid, and then found small molecular changes which might be harmful. 'There was no link between the experiment and everyday microwave warming,' he said.

Confusion has also been created over the dangers of fat and dairy products in the average diet, the link between coffee and infertility, the fear over minute particles of benzine in Perrier water, and whether beef was safe to eat or not. 'Red meat is a fantastic source of healthy nutrients and a fantastic source of pleasure,' Mr Bazalgette said.

Campaigners on food safety said that although some stories might be exaggerated it was important that the public was kept as informed as possible about the risks of some foods.

'Paul Bazalgette is not a scientist and he has no idea what he is talking about,' said Richard Lacey, professor of medical microbiology at the University of Leeds, who has undertaken research on the safety of beef and eggs.

'There are a number of people who want food safety to be improved and who believe that the public has a right to know the latest information. They should not be attacked for that.'

© The Guardian December, 1997

33

Why dicing with death isn't on the menu

Robert Uhlig weighs up the percentage risk of contracting CJD by eating a T-bone steak against other more dangerous everyday activities

It is a suspected killer that has caused untold misery in farming communities, threatened a centuries-old industry, prompted demonstrations at ports and initiated a £4 billion campaign to stamp it out, yet it is less dangerous than taking a bath.

There is a one in 20 chance of one person in Britain contracting new-variant CJD as a result of eating beef on the bone in the next year, according to Seac, the Government's advisers on the various forms of spongiform disease including BSE and CJD. Based on Meat and Livestock Commission figures that half the population eats beef, the risk of death faced by each of us through eating beef on the bone during the coming year is around one in 600 million.

Dr Ian Langford, a senior research fellow in risk perception and statistics at the University of East Anglia, said the figure was 'a long way beyond what we would call even a negligible risk'.

Nevertheless, this minuscule hazard is enough to prompt the Government to ban T-bone steaks, roast ribs of beef and oxtail.

The trivial scale of danger is made apparent by the Chief Medical Officer's scale of risk. According to this measure, there is a greater chance of dying from playing football, soaking in the bath, or doing just about anything else.

'At the top of the list, at scale 10, is life,' said Dr Langford. 'If you live, there is a one in one chance that one day you will die. At nine, that's a one in 10 risk, is the chance of contracting cancer at some point in your life, although that doesn't mean you will die from it.'

The most life-threatening activity on the scale is smoking 10 cigarettes a day, which has a one in 200 chance of causing death in a year – between seven and eight on the scale of risk.

Further down the scale are death from violence or poisoning (one in 3,300), from influenza (5,000), road accident (8,000), leukaemia (12,500), playing soccer (25,000), accident at home (26,000), accident at work (43,500), murder (100,000), railway accident (500,000), drowning in bath (800,000), or lightning (10 million). All these figures are giants in comparison with the minuscule risk of death from eating beef on the bone; a risk so small that it almost slips off the scale.

Death by T-bone steak appears at 1.5, several places below where risk is considered to be negligible, and a long way below such unlikely events as being hit by a meteorite (one in 10,000), in an air crash (one in 20,000) or being hit by a falling aircraft (one in 20 million). This highlights the incongruous nature of statistics.

Though nobody in living memory has been hit by a meteorite, when it does happen, millions of people will be killed at once. Plane crashes are more common, but kill fewer people over time, hence the lower risk of dying in an air crash.

Despite the apparent low risk of eating beef on the bone, Dr Langford said there was a large degree of uncertainty associated with the Government's estimate.

He said: 'We are at a very early stage in the epidemic and we don't even know if there are several forms of human CJD.'

© *Telegraph Group Limited, London 1997*

The beef crisis

The chances of any one individual dying in any one year from . . .

. . . contracting CJD by eating beef on the bone **1 in 600 million**	. . . being struck by lightning **1 in 10 million**	. . . drowning in the bath **1 in 800,000**	. . . homicide **1 in 100,000**	. . . playing soccer **1 in 25,000**
. . . a plane crash **1 in 20,000**	. . . being involved in a road accident **1 in 8,000**	. . . influenza **1 in 5,000**	. . . smoking 10 cigarettes a day **1 in 200**	The chances of winning the jackpot in the National Lottery **1 in 14 million**

Source: University of East Anglia

'Action needed' to stop food poisoning

The Government is not acting quickly enough to curb outbreaks of potentially fatal food poisoning, according to a leading food safety expert. Glenda Cooper, Consumer Affairs Correspondent, reports

Professor Hugh Pennington, who carried out the investigation into the E. Coli outbreak in Scotland which killed 20 people, said yesterday that his recommendations were not being carried out quickly enough. The E. Coli outbreak happened in November 1996 and Professor Pennington published his report in April 1997.

A record number of people suffered from food poisoning last year – 100,000 cases were reported officially but scientists estimate the real number could be 10 times that figure.

The criticism comes the day before the Government publishes the long-awaited White Paper on the new, independent Food Standards Agency. It also follows warnings from the British Medical Association that all raw meat should be treated as potentially contaminated.

Speaking on BBC2's *Food & Drink* programme, to be screened tonight, the professor said the figures were 'unacceptable' for a disease which was completely preventable.

'There's a crisis in British food production – it's not about BSE or healthy eating – it's food poisoning,' he said. 'If [the Government] are really serious they must implement all the recommendations in my report now... We need to set our standards higher – it would only take a few simple steps to get rid of the majority of food poisoning cases.'

Measures he said must be implemented without delay included licensing for butchers and restaurants, less reliance by supermarkets on intensively farmed foods and better training for people handling food.

'Food poisoning in the UK has now reached unacceptable levels,' he said. 'A million cases a year is outrageous.'

'The tragedy is that most cases of food poisoning are preventable – but they're not being prevented. It is an unnecessary problem,' he added. 'Education and training is the key. There are too many unqualified people handling food at each stage of the food chain.

'They need to be better qualified – it is, after all, a life and death issue.'

Butchers reacted angrily yesterday to the BMA's recommendation that all raw meat should be treated as potentially contaminated.

The Meat and Livestock Commission accused doctors of 'scaremongering' and exaggerating the dangers associated with meat.

'All fresh food is perishable and should be regarded as a possible source of contamination and red meat is no different to any other raw food requiring cooking before eating,' said Colin Maclean, director of the MLC.

But he said singling out meat as a possible hazard was likely to 'frighten and confuse' consumers.

*© The Independent
January, 1998*

The CJD toll

The numbers of deaths of definite and probable cases of Creutzfeld-Jakob disease in the UK.

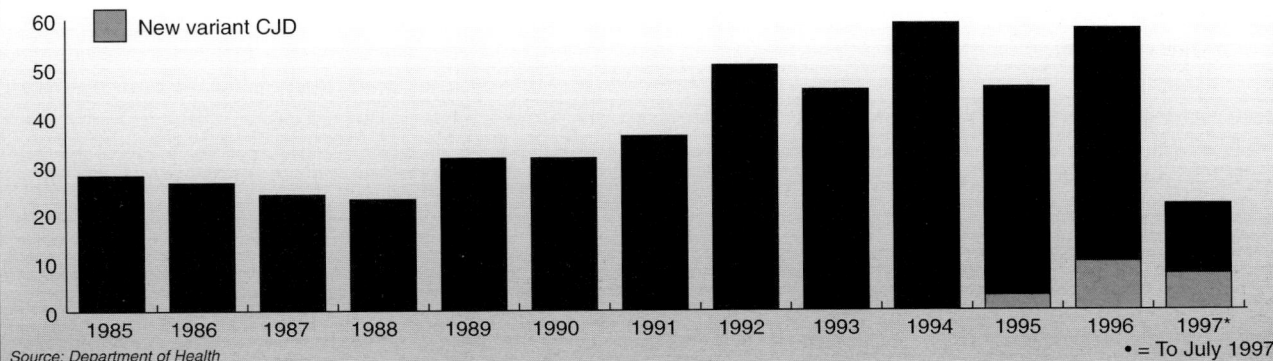

New variant CJD

Source: Department of Health
• = To July 1997

Foul food

Can the Government protect us from killer bugs?

There have been nearly 200 cases of serious food poisoning per day over the past decade. In the wake of BSE, E. Coli and salmonella, the Government yesterday laid out its plans for a food-safety agency. Glenda Cooper, Consumer Affairs Correspondent, examines how ministers plan to restore consumer faith in food.

The Government wants every shop and restaurant in Britain to pay £100 a year to finance the Food Standards Agency, which it promises will protect consumer interests in every area of food safety.

But farmers and food industry lobbyists warned that forcing the industry to pay for the agency would amount to a 'tax on food by the back door'.

The creation of one of the most powerful food watchdogs in Europe comes after a loss of public confidence in food safety after fears over the link between 'mad cow' disease and Creutzfeldt Jakob disease, the E. Coli 0157 outbreak in Scotland which killed twenty people and the scare over baby milk and 'gender bender' chemicals.

The FSA will be responsible for ensuring that food is safe to eat, and for advising people on what makes a healthy diet.

The Scottish health minister, Sam Galbraith, said that one of the main catalysts for the White Paper had been the E. Coli outbreak in Lanarkshire. 'From the plough to the plate, the agency will put consumers first,' he said. 'It will have tough powers to make sure the high standards we are aiming for are met all the way from farms to shops, from restaurants and to our kitchens.'

The agency will also co-ordinate food law enforcement and commission research, taking over many of MAFF's roles. It will also have a key role in nutrition, identifying and recommending balanced and nutritious diets for the public, although Jack Cunningham, the agriculture minister, said that it would not mean the agency would be telling people what they should eat. The agency would, however, be 'radical' and bring about 'fundamental changes', he added.

Part of it will be paid for by the food industry itself which the White Paper says 'should bear the bulk of the costs of improving food safety and standards' as the industry 'will benefit from the improved public confidence in safety'.

The suggestion is that the 600,000 shops, restaurants and manufacturing plants which are already registered with local authorities would have to pay for a licensing fee. The suggested flat rate of £100 per premises per annum

---

A number of people suffered from food poisoning last year – 100,000 cases. But scientists estimate the real number could be ten times that figure

---

would raise about £60m, 'a substantial amount'. It is expected that the agency's total annual expenditure will be more than £100m.

Frank Dobson, the Secretary of State for Health, stressed yesterday that that would not give the food industry power over the agency. 'Industry will not be laying down conditions,' he said.

Geoff Rooker, food safety minister, added: 'There will be no "no go areas" as far as we are concerned. It will be an independent agency.'

But the proposals in the White Paper were met with dismay by the food industry lobbyists who had argued against the inclusion of nutritional advice and have expressed opposition to any suggestion that businesses should pay for the agency's work.

'The Food Standards Agency should not jeopardise its independence through being funded by the food industry,' Sir David Naish, president of the National Farmers' Union, said. 'Its remit includes food safety which is clearly a public health matter which should be funded by the public purse.'

Doctors, consumer associations and public health experts welcomed the White Paper.

The agency, which will report to the Department of Health, not MAFF, will be made up of a commission of twelve independent people backed up by advisory committees and 'several hundred' civil servants.

Earlier this week it was reported that a record number of people suffered from food poisoning last year – 100,000 cases. But scientists estimate the real number could be ten times that figure.

© The Independent
January, 1998

Stop bugging me!

Why food poisoning is on the increase and how you can protect yourself and your family. By Professor Richard Lacey

What is food poisoning?

It is people being poisoned by food that they have eaten – or rather by certain types of bugs that are in it. These bugs – called bacteria and viruses – are so small they cannot be seen by the eye. They may grow in food and cause you to become sick. Sometimes, people are not poisoned by the bugs themselves, but by chemicals they release into the food that is eaten. Nice eh?

What does food poisoning do to you?

Have you ever experienced this? About 12 hours after eating chicken you start to feel sick. This grows into sharp pains that keep shooting from your stomach to your back. Then you get explosive diarrhoea – you can't hold it back and are never off the loo. Now you feel really ill! You feel hot and start to vomit. This goes on for a few days and then you feel tired out for a couple of weeks. You swear you will never eat chicken again! That is what a typical bout of salmonella food poisoning does to you.

Who is at risk?

You are – particularly if you eat meat, fish or eggs. Figures based on the Government's Communicable Disease Surveillance Centre show that a massive 700,000 people in Britain suffered from food poisoning in 1991.

More than 90% of these cases were caused by eating animals or animal products.

Some people are more prone to suffer the effects of food poisoning than others. The elderly, pregnant women, newborn babies and people who are already ill may even die from food poisoning. In Britain in 1991, one hundred people died from salmonella poisoning and 50 from another bacteria called E. coli, found in foods such as burgers and sausages.

And things are not getting better. Food poisoning increased fourfold in the UK between 1982 and 1992.

Why is eating animal products a big cause of food poisoning?

It makes sense that bugs in animals are more likely to infect us than bugs found in plants. Animals are closer to us biologically than plants and so, for example, bacteria in a cow are more adapted to living in cells that are similar to human cells, rather than bacteria from a carrot.

In fact, people catch many diseases from cattle including tuberculosis, listeria, cryptospiridium, salmonella, E. coli poisoning and possibly BSE or mad cow disease. We can also be infected by certain types of bugs from all the other kinds of farm animals.

One of the major reasons that food poisoning has increased so rapidly in the last couple of decades is that farm animals are increasingly subjected to intensive or factory farming. This often means confining them in small spaces throughout their lives, pushing them beyond their limits simply so they grow faster, so that they can be killed earlier.

The broiler chicken is a prime example. Broilers are the chickens that are eaten (they are not used for egg production). Almost all broiler chickens in Britain are 'farmed' by being crammed into sheds in their

WAITER! THERE'S AN ENVIRONMENTAL HEALTH OFFICER IN MY SOUP!

Ken Pyne

thousands where they stand in their own excrement.

They are genetically bred to grow quickly but in order to make them grow even faster they are given growth-promoting drugs. The bones of these chickens break under their ballooning weight and their hearts cannot cope. About 33 million die in these factory farms each year. Bacteria such as salmonella and campylobacter inevitably abound in the thousands of tons of droppings and rotting dead carcasses and infect tens of thousands of birds.

Battery hens are stacked in tiny cages in dim, stinking sheds. After an egg is laid it rolls into a collecting gulley. Feed and water are supplied automatically and lights are on for about 17 hours a day to promote egg laying. Up to 30,000 birds are in these sheds. The combination of a lack of fresh air, selective breeding and the caging of the birds in overcrowded conditions so that they cannot even exercise, has led to the spreading of disease and to distress and suffering.

Both broiler and battery chickens are often fed the infected remains of their own kind – allowing salmonella to spread like wild-fire. The bug can get into the hen's ovaries and from there into her eggs. If you eat an infected egg that is not properly cooked, the salmonella may grow in you! In the case of broilers, 50 to 80% are infected with salmonella. Again, if you eat a bird that is not thoroughly cooked, you could be heading for trouble.

It has been known for years that modern intensive methods of broiler and egg production are addled with salmonella. The UK Junior Minister of Health, Edwina Currie, was right when she said in 1988 that most of our egg production was contaminated with salmonella.

In my laboratory in Leeds, we are identifying more food poisoning bacteria from ill patients then ever before. The problem is not solved but is getting worse.

All factory-farmed animals whether they are hens, pigs, cows or turkeys are treated as commercial commodities and not as living, feeling creatures that suffer pain and fear in much the same way as you or

---

It has been known for years that modern intensive methods of broiler and egg production are addled with salmonella

---

me. Because this type of farming is motivated entirely by the desire for profit, animal welfare and safety are not considered. The farmer – or should we say 'investor' – is interested only in ways by which the animal can gain weight over the shortest period of time for the lowest cost.

He will take the minimum precautions needed to control disease and has been generally unconcerned over the exact contents of the animal feed, so long as it does its job – putting on weight. This is at the heart of the dilemma today.

How slaughter spreads disease

The slaughterhouse is responsible for spreading bugs on a massive scale.

Taking chickens as an example, after being killed, they are plucked and have their guts removed on a conveyor belt system where water and knives spread bacteria from one carcass to the next. The same type of thing happens to cows, pigs and sheep – namely that the lack of hygiene and space spreads bugs all over the place. If a living animal went into a slaughterhouse bug-free, there is a good chance that it would emerge as a disease-ridden carcass.

Other examples of food poisoning

Campylobacter

Campylobacter is the ugly name of the ugly bug that causes most food poisoning in Britain. 350,000 people are stricken with it each year. It is mainly caused by eating infected chicken or turkey that is under-cooked, or from unpasteurised milk.

As with salmonella, it can be spread from contaminated raw chicken to other foods, for example by using the same knife to cut your veg as your chicken. This bug makes you want to vomit, gives you stomach ache, sharp, nasty pains and a lot of diarrhoea. The illness lasts for 5-10 days, with a further 1-2 weeks before you feel well again.

Eating animals is a big risk in terms of food poisoning and is making almost a million people ill – some very seriously – every year. 95% of all food poisoning cases are caused by eating animal products.

A tiny 5% are from plant foods and much of this is from contamination by animal manure or from animal products. A diet free of meat and fish and, if you are vegan, of dairy products and eggs, is by far the safest and one that I highly recommend.

• Richard Lacey is Professor of Clinical Microbiology at Leeds University. He has a degree in medicine from Cambridge University and a Ph.D in clinical microbiology from the Faculty of Medicine at the University of Bristol.

He acts as consultant to the World Health Organisation and was a member of the Veterinary Products Committee of the government's Ministry of Agriculture from 1986 to 1989.

He has received numerous awards, including the Evian Health Prize for Medicine, the Caroline Walker Award for Science and an award from the Campaign for Freedom of Information.

• The above is an extract from *Stop bugging me!*, by Professor Richard Lacey, it is the second guide in a series produced by Viva! See page 41 for address details.

© Viva! – Vegetarians International Voice for Animals

New beef rules baffle butchers

Meat row: Regulations imprecise, say officials. By Owen Bowcott

The Government's ban on selling beef on the bone came into force at midnight on Monday. By breakfast yesterday the hastily-drafted regulations were already being branded 'unenforceable'.

Restaurant owners, butchers, farmers and caterers were all claiming that the Beef Bones Regulations 1997 – passed as statutory number 2959 – left them bewildered about what meat and stocks could still be used for human consumption.

As more than 1,000 Scottish and Welsh beef, lamb, pig and cereal farmers arrived in Westminster to protest at falling agricultural incomes triggered by the BSE crisis, even officials at the Ministry of Agriculture, Fisheries and Food (MAFF) admitted that the regulations were an 'imprecise science'. One veteran butcher, Ray Robinson, of Burntwood, Walsall, West Midlands, warned that he was prepared to risk prosecution rather than stop selling T-bone steaks and other cuts on the bone.

'I have already sold T-bone steaks,' he said yesterday, 'and people are still asking for those cuts.'

The main doubts yesterday centred on fresh meat.

Is all beef on the bone now banned?
Not quite. T-bone steaks, oxtail, sirloin on the bone, fore-ribs and roast rib of beef are forbidden fare – unless the animal was under six months old.

How will anyone know the exact age of a carcass?
The Meat and Livestock Commission says that if not sold as veal calves, cattle are normally slaughtered only when much older. Each animal is supposed to have a passport document identifying its age. MAFF suggests that inspectors will know by judging the size of any joint, but admits that 'it may be an imprecise science'.

Who will enforce the regulations?
Environmental health officers and trading standards inspectors in local councils. An official of the Chartered Institute of Environmental Health yesterday told a Commons committee, however, that his organisation had not been consulted and that the regulations may not be enforceable.

Are all soups, stock and gravy made from meat on the bone banned?
Yes. Butchers will not even be allowed to give away bones to people claiming to want them for their dogs.

When must bones be removed?
Butchers can sell meat on the bone to restaurants but not consumers. Restaurants have to remove the bones before cooking. So meat could be cut off the bone as little as 20 minutes before it reached the diner's plate.

What about stock cubes?
Manufacturers have been given three months to clear the shelves of products made with British beef bones. Most household names, however, such as Oxo and Bovril, already source their products from BSE-free countries abroad.

What about the meat stored in freezers?
Government advice is do not eat it – but it is not against the law. Defiant consumers have been stockpiling supplies against future shortages.

Who will be prosecuted?
Anyone found selling beef on the bone is liable to a £5,000 fine and six months' imprisonment.

Any customer persuading his butcher or restaurant to sell meat on the bone could, conceivably, also face charges.

Is there anything to stop a trade in beef on the bone from cross-channel customers shopping in France?
No. Beef imports for 'personal use' will not be blocked.

This has prompted speculation that T-bone steak runs could become as popular as booze cruises.

© *The Guardian* December, 1997

Sorry, Toby, no more bones.

£300m meat bug bill

From farm to shop, an all-out war on E. Coli. By Sean Poulter, Consumer Affairs Correspondent

The meat industry faces a near £300 million bill to improve hygiene standards in the wake of the E. Coli outbreaks which left 20 dead and hundreds seriously ill.

A shake-up in food safety laws, covering everything from ensuring animals are clean at slaughter to hygiene standards in butchers' shops and supermarkets, was recommended by the inquiry into the food poisoning tragedy in Scotland yesterday.

Professor Hugh Pennington's key proposals include a new licensing system for food premises and the creation of a 'hygiene wall' between the handling and sale of raw and unwrapped cooked meats.

Shops will have to find the money for new utensils, storage facilities and extra staff to end the risk of cross-contamination.

Publication of the report sparked a fierce political row as opposition parties accused the Government of ignoring earlier warnings about lax hygiene standards at slaughterhouses, seen as a key factor in the outbreaks.

E.Coli 0157 is an extremely virulent organism. A tiny dose can cause serious illness or death, particularly among vulnerable groups such as the elderly and the young.

Professor Pennington, of Aberdeen University, was asked to conduct his inquiry following the Lanarkshire outbreak last year. He also investigated a further outbreak in Tayside.

He said training and education are needed to ensure good hygiene standards are adopted by everybody involved in meat handling, from the farm to the supermarket shelf.

Scottish Secretary Michael Forsyth said ministers accepted all 30 of the inquiry's recommendations. He added that steps had already been taken to raise standards in meat plants, but more action is needed.

This will include a sustained publicity campaign to alert the industry and the public to the dangers.

Labour accused the Government of relaxing food safety standards in the past and putting consumers at risk.

The party's Scottish spokesman, George Robertson, said: 'Their handling of the E.Coli outbreak and their failure to take adequate preventative action are just the latest in a long line of failures which have put lives at risk.

'The message of this report is that if the public are going to be reassured about the food that they buy and the food that they eat, there must be an independent food standards agency.'

Liberal Democrat spokesman Paul Tyler joined the call for a new body to protect consumer interests and restore public confidence.

The Consumers' Association also called for an independent agency and said the recommendations did not go far enough. Director Sheila McKechnie said: 'There is still no mechanism to ensure the recommendations are followed through.'

Stores large and small were last night considering their reaction to the report. Tesco said it would review the findings and was considering new ways of segregating meats in stores.

The National Federation of Meat and Food Traders, which represents about 3,500 independent retail butchers, said that licensing went too far. The costs involved could force some shops out of business.

Shops will have to find the money for new utensils, storage facilities and extra staff

---

Farms

The main source of E.Coli 0157 is the intestines and faeces of cattle and, possibly, sheep.

The report calls for immediate action to bring down contamination levels on farms and minimise the risk of the organism getting into food.

The key recommendation requires the Meat Hygiene Service, which is responsible for monitoring abattoir standards, to 'rigorously enforce' rules demanding that animals sent for slaughter are not dirty.

Abattoirs

Hygiene failures in this area are responsible for the deadly organism reaching butchers, the report makes clear.

It calls for a system of checks – the Hazard Analysis and Critical Control Point system – to be used in all 400 slaughterhouses and abattoirs to identify dangers.

Professor Pennington also insists that more care is needed to ensure that intestines and hides are removed in a way that minimises the risk of contamination.

Shops

Licensing will be policed by health officers with the power to close premises.

The report calls for the separation, in storage, production, sale and display, between raw meat and unwrapped cooked meat. This should include the use of separate fridges and production equipment, utensils and, wherever possible, staff.

Where not possible, washing facilities must be installed in the serving area to deal with the risk of cross-contamination.

ADDITIONAL RESOURCES

You might like to contact the following organisations for further information. Due to the increasing cost of postage, many organisations cannot respond to enquiries unless they receive a stamped, addressed envelope.

British Nutrition Foundation (BNF)
High Holborn House
52-54 High Holborn
London, WC1V 6RQ
Tel: 0171 404 6504
Fax: 0171 404 6747
The (BNF) is an independent charity which provides reliable information and advice on nutrition and related health matters. They produce a wide range of information. Ask for their publications list.

European Food Information Council (EUFIC)
1 Place des Pyramides 75001
Paris
France
Tel: 00 33 140 20 44 40
Fax: 00 33 140 20 44 41
Web site: See page 43 for details
EUFIC is a non-profit making organisation based in Paris. It has been established to provide science-based information on foods and food-related issues. It publishes regular newsletters, leaflets, reviews, and other background information on food issues.

Meat & Livestock Commission (MLC)
PO Box 44, Winterhill House
Snowdon Drive
Milton Keynes, MK6 1AX
Tel: 01908 677577
Fax: 01908 609826
Answers consumers' concerns on meat eating, diet and health, and the importance of red meat in a healthy, balanced diet.

Ministry of Agriculture, Fisheries and Food (MAFF)
Publications Department
London, SE99 7TP
Tel: 0645 556000
Publish *Genetically Modified Food* (ask for Ref: 2052) in their Food Sense series. Publications are free but take up to ten working days for delivery.

National Farmers' Union (NFU)
164 Shaftesbury Avenue
London, EC2A 2BH
Tel: 0171 331 7200
Fax: 0171 331 7313
The NFU takes a close interest in the whole range of rural affairs and works with politicians and officials – both in the UK and internationally – and other groups and organisations to advance rural interests.

National Food Alliance Project (NFA)
94 White Lion Street
London, EC2A 2BH
Tel: 0171 837 1228
Fax: 0171 837 1141
E-mail: nationalfood alliance@compuserve.com
Represents national public interest organisations including voluntary, professional, health, consumer and environmental bodies working at international, national, regional and community level on food related issues. Produces newsletters, books etc.

The Food Commission
94 White Lion Street
London, N1 9PF
Tel: 0171 837 2250
Fax: 0171 837 1141
Provides education, information, advice and research on nutrition, diet, health and food production. Runs various educational and research campaigns, publishes *Food Magazine* and other publications. Ask for their publications list.

The Soil Association
Bristol House
Victoria Street
Bristol, BS1 6DF
Tel: 0117 929 0661
Fax: 0117 925 2504
Works to educate the general public about organic agriculture, gardening and food, and their benefits for human health and the environment.

The Vegan Society
Donald Watson House
7 Battle Road
St. Leonards-on-Sea
East Sussex, TN37 7AA
Tel: 01424 427393
Fax: 01424 717064
Works to promote a way of life entirely free from animal products for the benefit of humans, animals and the environment. Encourages the development and use of alternatives to all commodities normally derived wholly or partly from animals. The society produces magazines, factsheets and a wide range of other literature. Ask for their publications list.

The Vegetarian Society of the United Kingdom Ltd
Parkdale
Dunham Road
Altrincham
Cheshire, WA14 4QG
Tel: 0161 928 0793
Fax: 0161 928 0893
Works to increase the number of vegetarians in the UK in order to save animals, benefit human health and protect the environment and world food resources. Publishes a variety of information on vegetarianism and also has a mail order book service. Ask for their publications list.

VIVA!
12 Queen Square
Brighton, BN1 3FP
Tel: 01273 777688
Fax: 01273 776755
Publish 12 Viva! guides on vegetarianism. They also have a new book published in September 1997: *The Livewire Guide to Going, Being and Staying Veggie* by Juliet Gellatley. Published by the Women's Press.

INDEX

Independence Web News

Back	Forward	Home	Reload	Images	Open	Print	Find	Stop

Live Home Page	Net Search	Apple Computer	Apple Support	Apple Software

The Internet has been likened to shopping in a supermarket without aisles. The press of a button on a Web browser can bring up thousands of sites but working your way though them to find what you want can involve long and frustrating on-line searches. And unfortunately many sites contain inaccurate, misleading or heavily biased information. Our researchers have therefore undertaken an extensive analysis to bring you a selection of quality Web site addresses. If our readers feel that this new innovation in the series is useful, we plan to provide a more extensive Web site section in each new book in the *Issues* series.

* * * * *

Centre for Advanced Food Research
Australia
http://hotel.uws.edu.au/~geoffs/
Databases include code breaker for food additives, recommended daily allowances of nutrients for adults, dietary guidelines for people of all ages, and physiology of sports and isotonic drinks. Useful list of sources for Food, Nutrition and Health information.

European Food Information Council (EUFIC)
http://www.eufic.org
As an independent, credible and knowledgeable source, EUFIC ensures the flow of clear, relevant and reliable information to consumers. Provides information on food-related issues in order to inform consumers on nutritional quality and safety of foods.

Health Education Authority (UK)
http://www.hea.org.uk
The role of the Health Education Authority (HEA) is explained. The web site serves as a gateway to various HEA health promotions and campaigns. Links are also provided for research and information resources like the European Network of Health Promoting Schools (ENHPS).

Mayo Foundation for Medical Education and Research
http://www.mayohealth.org/mayo/ 9511/htm/ecoli.htm
General information on protecting yourself against E. coli bacteria. Signs and symptoms of food-borne illness and preventative measures. Has a one-page clinical brief on E. coli, with a few facts and figures on incidences of contaminated food.

Think Fast
http://www.thinkfast.co.uk
Their Fact File provides information about popular fast foods, so one can choose healthier fast food options. Under scrutiny are burgers, chicken, kebabs, fish and chips, pizzas, pasta, sandwiches and other staple foods on the UK market. Quiz show and other health food campaigns are available through links from this site. Visitors can send friends a virtual fast food postcard.

World Health Organization (WHO)
http://www.who.ch/programmes/emc/bsefacts.htm
Reports on Emerging and Other Communicable Diseases (EMC). Fact sheet (March '96) on Bovine Spongiform Encephalopathy (BSE). Information provided on Creutzfeldt-Jakob disease (CJD). Has an informative brief on BSE and CJD.

ACKNOWLEDGEMENTS

The publisher is grateful for permission to reproduce the following material.

While every care has been taken to trace and acknowledge copyright, the publisher tenders its apology for any accidental infringement or where copyright has proved untraceable. The publisher would be pleased to come to a suitable arrangement in any such case with the rightful owner.

Chapter One: The Food we Eat

Children's views on food and nutrition, © EUFIC, October 1995, *Growing up green*, © The Guardian, January 1997, *Lunch*, © The University of Birmingham, *Britons chill out in kitchen*, © The Guardian, September 1997, *10 tips to healthy eating*, © EUFIC, *Current eating and drinking patterns*, © EUFIC, October 1995, *Children's views on nutrition and health*, © EUFIC, *Chips are down for the canteen*, © Telegraph Group Limited, London 1997, *Food for thought*, © DfEE, *Junk food is taken off the menu*, © Telegraph Group Limited, London 1997, *No jacket potato required?*, © Direct Response, September 1997, *Lunching habits*, © Eurest, *Grazing gets a grip and Britain becomes a nation of snackers*, © Key Note, *Britons the 'burger kings of Europe'*, © The Daily Mail, 1997, *Italy succumbs to English taste*, © Telegraph Group Limited, London 1997, *Can't cook, won't cook, say women*, © The Daily Mail, November 1997, *Wales the target in campaign to promote healthier eating habits*, © The Guardian, June 1997, *Understanding obesity*, © EUFIC, *Slimmers' yoghurt claims to make stomach feel full*, © The Independent, January 1998, *What the label doesn't tell you*, © Food Magazine, February 1998, *Your right to know*, © Food Magazine, February 1998, *Shoppers 'confused' by nutrition labels on food*, © The Guardian, June 1997, *What to eat*, © The Guardian, June 1997, *Understanding food*, © European Food Information Council (EUFIC).

Chapter Two: Food Safety

Foodborne illness, © EUFIC, *New food watchdog could cost shoppers an extra £14 a year*, © The Daily Mail, January 1998, *Food safety and your baby*, © Merton Environmental Health Department, *Danger labels are slapped on eggs, cheese, carrots, baby milk – and meat*, © The Independent, January 1998, *E numbers*, © EUFIC, *Eating outside the home*, © EUFIC, *Causes of foodborne disease in service establishments*, WHO Europe Surveillance Reports, 1990-1992, *Food poisoning*, © Merton Environmental Health Department, *Food poisoning*, © EUFIC, *Beef: a crisis out of control*, © The Guardian, December 1997, *Doctors warn that meat is dangerous*, © The Independent, January 1998, *Food industry backlash over 'lunatic' measures*, © The Daily Mail, December 1997, *TV food executive hits out over scares*, © The Guardian, December 1997, *Why dicing with death isn't on the menu*, © Telegraph Group Limited, London 1997, *The beef crisis*, University of East Anglia, *'Action needed' to stop food poisoning*, © The Independent, January 1998, *The CJD toll*, © Department of Health, *Foul food*, © The Independent, January 1998, *Stop bugging me!*, © VIVA!, *New beef rules baffle butchers*, © The Guardian, December 1997, *£300 m meat bug bill*, © The Daily Mail, April 1997.

Photographs and illustrations:

Pages 1, 18, 23, 33, 39: The Attic Publishing Co, pages 4, 6, 17, 21, 30, 37: Ken Pyne.

Thank you

Darin Jewell for assisting in the editorial research for this publication.

Craig Donnellan
Cambridge
April, 1998

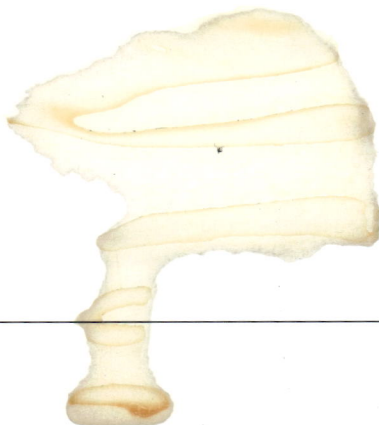